Surveys and Tutorials in the Applied Mathematical Sciences

Volume 3

Series Editors
S.S. Antman, J.E. Marsden, L. Sirovich

Surveys and Tutorials in the Applied Mathematical Sciences
Volume 3

Series Editors

S.S. Antman, Department of Mathematics and Institute for
Physical Science and Technology, University of Maryland,
College Park, MD 20742-4015, USA; ssa@math.umd.edu

J.E. Marsden, Control and Dynamical System, 107-81,
California Institute of Technology, Pasadena, CA 91125, USA;
marsden@cds.caltech.edu

L. Sirovich, Laboratory of Applied Mathematics, Department of
Bio-Mathematical Sciences, Mount Sinai School of Medicine,
New York, NY 10029-6574, USA; Lawrence.Sirovich@mssm.edu

Mathematics is becoming increasingly interdisciplinary and developing stronger interactions with fields such as biology, the physical sciences, and engineering. The rapid pace and development of the research frontiers has raised the need for new kinds of publications: short, up-to-date, readable tutorials and surveys on topics covering the breadth of the applied mathematical sciences. The volumes in this series are written in a style accessible to researchers, professionals, and graduate students in the sciences and engineering. They can serve as introductions to recent and emerging subject areas and as advanced teaching aids at universities. In particular, this series provides an outlet for material less formally presented and more anticipatory of needs than finished texts or monographs, yet of immediate interest because of the novelty of their treatments of applications, or of the mathematics being developed in the context of exciting applications. The series will often serve as an intermediate stage of publication of material which, through exposure here, will be further developed and refined to appear later in one of Springer's more formal series in applied mathematics.

Surveys and Tutorials in the Applied Mathematical Sciences

Thomas Erneux

Applied Delay Differential Equations

Thomas Erneux
Lab. Optique Nonlinéaire Théorique
Faculté des Sciences
Campus de la Plaine
Université Libre de Bruxelles
bd. du Triomphe
1050 Bruxelles, Belgium
terneux@ulb.ac.be

ISBN 978-0-387-74371-4   e-ISBN 978-0-387-74372-1
DOI 10.1007/978-0-387-74372-1

Library of Congress Control Number: 2008939990

Printed on acid-free paper

Springer Science+Business Media
springer.com

Preface

Qu'est-ce que le passé, sinon du présent qui est en retard?
Pierre Dac (l'os à moelle, March 1940)

Modeling automatic engines or physiological systems often involves the idea of control because feedback is used in order to maintain a stable state. But much of this feedback require a finite time to sense information and react to it. A popular way to describe this process is to formulate a delay differential equation (DDE) where the evolution of a dependent variable at time t depends on its value at time $t - \tau$. Unfortunately, solving a DDE is a mathematically difficult task. Over the past decade, rapid advances in computational power have revived interest in DDEs. Previously known equations are investigated allowing a better physical understanding of old problems. In addition, new areas of research have appeared. This is, for example, the case of lasers subject to optical feedback, the delayed control of container cranes, or the real-time synthesis of musical instruments.

Oscillatory instabilities are frequently associated with systems described by DDEs. The motivation to study these oscillations then depends on the background of the researcher. For some, these oscillatory instabilities are viewed as a limitation to the performance of a particular device that must be avoided or possibly controlled. In contrast, other researchers have put the unstable behavior to good use making practical devices such as high-frequency optical oscillators.

New mathematical tools and reliable computer software techniques have been developed for DDEs. Here, preference is given to analytical approaches known collectively as asymptotic methods [22, 124], the most useful techniques for finding approximate solutions to equations. It is a revised and largely expanded version of a series of lectures first given at the Université Libre de Bruxelles in 2002–2003, at the Université Joseph Fourier (Grenoble) in 2003, and, more recently, at the University of Utah in 2007. The minimum prerequisites for this book are a facility with calculus, experience with differential equations, and an elementary knowledge of bifurcation theory. The unusual format of this textbook, avoiding rigorous mathematical proofs and concentrating on applications, aims to introduce beginning students as well as experienced researchers to the large variety of phenomena described by DDEs. It has no ambition to review the rich field of DDEs and references have been selected for their historical impact or for the experiments they are describing.

One novelty in this volume is the place given to the figures. They help in understanding the scientific background of a specific application and how a mathematical model is derived. In addition, computer plots compare exact and approximative solutions illustrating the efficiency of the analytical method. The mathematical computations are described in as friendly a manner as possible.

DDE models are used by biologists, physicists, and engineers with different objectives and expectations. This text is meant to serve as an introduction to the rich variety of applications and could be used in a modeling course on DDEs. Some selected parts of this book could provide material for a class on singular perturbation techniques or to stimulate a differential equation class. It is my experience that the combination of illustrations using a projector and compact computations on the blackboard works well to attract the attention of the audience.

There are many colleagues to thank for their interest, suggestions, and contributions to this book. I first wish to thank Tamás Kalmár-Nagy and John Milton who gave me precious details on comparisons between experiments and theory. The collaboration with Dirk Roose and his group during the years 2000–2003 was a successful experience combining new analytical and numerical approaches. Applied mathematicians Don Cohen, Michael Mackey, and John Ockendon strongly encouraged me to go forward with this project. The lectures by Gabor Stépan on mechanical engineering problems, by Yang Kuang and Stephen Gourley on population models, and by John Mallet-Paret, Roger Nussbaum, and Hans-Otto Walther on state-dependent delay equations had a strong impact on me. Of course, I am deeply indebted to my friends in the laser community who in 1993 introduced me to the world of optical feedback. Tom Gavrielides, Vassilios Kovanis, and Daan Lenstra educated me on the complexities of the Lang and Kobayashi equations and Ingo Fisher, Eric Lacot, Laurent Larger, Raj Roy, and David Sukow patiently explained to me the subtleties of the experiments. Work could not have been done without the contribution of enthusiastic young collaborators: Kirk Green, Theodore Kolokolnikov, Michel Nizette, Didier Pieroux, Fabien Rogister, Marc Sciamanna, and Guy Van der Sande. I am grateful to Gregory Kozyreff for his constructive criticisms of the manuscript and I thank my editor Achi Dosanjh who convinced me to go for an ambitious project. I acknowledge the Belgian National Science Foundation and the Pole Attraction Pole program of the Belgian government (2001-2006) for the support I received during the preparation of this book. Lastly I would like to thank Anne, Joan, and Marc for their love and support while I was away searching and computing. This book is dedicated to them.

Brussels, Belgium   *Thomas Erneux*
January 2008

Contents

1

Introduction

After the First World War, the development and use of automatic control systems resulted in studies of an entirely different class of differential equations the so-called *delay differential equations* or *difference differential equations* (DDE). Any system involving a feedback control will almost certainly involve time delays. A time delay arises because a finite time is required to sense information and then react to it. Severe stability problems, however, appear as soon as several mechanisms need to be controlled simultaneously. The so-called pilot-induced oscillations (PIO), for example, consist of unintentional sustained oscillations resulting from efforts of the pilot to control the aircraft [168]. The books by Pinney [192] and Driver [52] contain a large compilation of DDEs that have appeared in the literature. The more recent books by MacDonald [146], Stépán [217], Kuang [132], Fowler [65, 67], Epstein and Pojman [55], Murray [172], Fall et al. [63], Beuter et al. [25], and Britton [34] emphasize particular DDE problems appearing in mechanical engineering, chemistry, and biology. Mathematically, the subsequent research of Mishkis [167], Bellman and Danskin [20], Bellman and Cooke [21], and Krasovskii [130] set the stage for the monumental work of Hale [84] and students at Brown. Advanced mathematical issues on DDEs, functional equations,[1] and robust control are treated in the books by Hale and Verduyn Lunel [85], Diekmann et al. [51], Kolmanovskii

---

[1]DDEs are a special class of more general *functional equations*, such as integrodifferential equations. In the restricted case of a DDE, only a certain finite interval of the immediate past is involved in the determination of the present.

T. Erneux, *Applied Delay Differential Equations*, Surveys and Tutorials in the Applied Mathematical Sciences 3, DOI 10.1007/978-0-387-74372-1_1,
© Springer Science+Business Media, LLC 2009

and Myshkis [119], and Michiels and Niculescu [161]. A large part of the revised "Handbook of Chaos Control" is devoted to delayed feedback control techniques [210]. Finally, reliable numerical methods have been developed for DDEs [227].

In this introduction, we emphasize two fundamental properties of a DDE and then propose a variety of examples.

1.1 Properties

A time-dependent solution of a DDE is not uniquely determined by its initial state at a given moment but, instead, the solution profile on an interval with length equal to the delay (or time lag) τ has to be given. That is, we need to define an infinite-dimensional set of initial conditions between $t = -\tau$ and $t = 0$. Thus, DDEs are infinite-dimensional problems, even if we have only a single linear DDE.

The simplest way to illustrate how a DDE differs from an ordinary differential equation (ODE) is to consider a linear first-order differential equation. We all know that the initial value problem

$$\frac{dy}{dt} = ky, \ y(0) = 1 \tag{1.1}$$

admits the exponential solution

$$y(t) = \exp(kt). \tag{1.2}$$

Physically, the knowledge of the present (here: $y(0) = 1$) allows us to predict the future at any time t. The past is not involved in this solution. For a DDE, the past exerts its influence on the present and, hence, on the future. The following DDE

$$\frac{dy}{dt} = ky(t - \tau), \quad y(t) = 1 \quad \text{when} \ -\tau \le t < 0 \tag{1.3}$$

exhibits a right hand side that depends on y at time $t - \tau$. τ is called the delay or time lag. Moreover, the initial condition is now replaced by an initial function defined on a finite interval of time. There are two important properties of this equation that need to be stressed.

1.1.1 Oscillations

In contrast to the exponential solution (1.2), the solution of Eq. (1.3) can be oscillatory. This can be seen by seeking a particular solution of the form

$$y = A\sin(\omega t). \tag{1.4}$$

Inserting (1.4) into Eq. (1.3), we find

$$
\begin{aligned}
\omega A \cos(\omega t) &= kA \sin(\omega t - \omega \tau) \\
&= kA \left[\sin(\omega t) \cos(\omega \tau) - \cos(\omega t) \sin(\omega \tau) \right].
\end{aligned}
\tag{1.5}
$$

Equating to zero the coefficients of $\cos(\omega t)$ and $\sin(\omega t)$, we find the following two conditions,

$$
\cos(\omega \tau) = 0 \quad \text{and} \quad \omega = -k \sin(\omega \tau).
\tag{1.6}
$$

The first condition is satisfied if $\omega \tau = \pi/2$ or $3\pi/2$ and with the second condition, we obtain the following possibilities,

$$
(1) \ \omega \tau = \pi/2 \quad \text{and} \quad k\tau = -\frac{\pi}{2},
\tag{1.7}
$$

$$
(2) \ \omega \tau = 3\pi/2 \quad \text{and} \quad k\tau = \frac{3\pi}{2}.
\tag{1.8}
$$

For these particular values of $k\tau$, the DDE (1.3) admits the harmonic solution (1.4). Condition (1.7) is used in Section 1.2 below.

1.1.2 Short time solution

The second and most obvious difference between ODEs and DDEs is the initial data. For DDEs we must provide not just the value of the solution at the initial point, but also the history, that is, the solution $y_0(t)$ at times prior to the initial point. We may analytically investigate the effect of $y_0(t)$ by using the *method of steps* [248]. For mathematical clarity, we consider the case $k = -1$, and $\tau = y_0 = 1$. We first solve Eq. (1.3) on the interval $[0, \tau[$. During this interval, Eq. (1.3) becomes

$$
\frac{dy}{dt} = -1
\tag{1.9}
$$

which we solve using $y(0) = 1$. The solution is

$$
y = 1 - t \quad \text{for } 0 \leq t < 1.
\tag{1.10}
$$

See Figure 1.1. Now that y is known up to 1, we consider the interval $[1, 2[$. Equation (1.3) becomes

$$
\frac{dy}{dt} = -1 + (t - 1)
\tag{1.11}
$$

with initial condition obtained from (1.10) at time $t = 1$; that is $y(1) = 0$. The solution is

$$
y = -(t - 1) + \frac{1}{2}(t - 1)^2 \quad \text{for } 1 \leq t < 2.
\tag{1.12}
$$

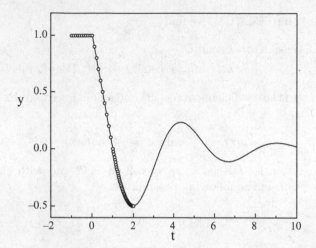

Figure 1.1: The method of steps is used for solving $y' = -y(t-1)$ with $y = 1$ during the time interval $[-1, 0[$. It provides the solution $y = 1 - t$ for $[0, 1[$ and $y = -(t-1) + \frac{1}{2}(t-1)^2$ for $[1, 2[$ (open circles).

See Figure 1.1. One can then consider the next interval $[2, 3[$, and so on. This procedure can, in principle, be continued as far as desired. But the calculations quickly become unwieldy without revealing essential properties of the solution[2]. As we show in the next chapter, there is a more interesting way to analyze the solution of Eq. (1.3). The method of steps is nevertheless valuable if we analyze the effect of $y_0(t)$ on a short time interval. In [16], the time history is analyzed in order to explain the staircase growth of a population of yeast cells.

Besides its effect on the short time behavior of the solution, the fact that we have an initial history has another impact when we numerically solve a DDE. Because numerical methods for both ODEs and DDEs are intended for problems with solutions that have several continuous derivatives, discontinuities in low-order derivatives require special attention. In our particular example, we note that

$$y'(0^-) = 0 \neq -1 = y'(0^+),$$

$$y''(1^-) = 0 \neq 1 = y''(1^+) \tag{1.13}$$

so that the jump in $y'(t)$ at $t = 0$ propagates to a jump in $y''(t)$ at $t = 1$, and so on. More generally, the jump in $y'(t)$ at time $t = 0$ propagates to

---

[2]Marc R. Roussel lecturing on DDEs proposed to automate these calculations using Maple. The damped oscillatory solution is then made up in piecewise fashion by a set of functions. See: http://people.uleth.ca/~roussel/.

a jump in $y^{n+1}(t)$ at time $t = n$. The propagation of discontinuities is a feature of DDEs that does not occur in ODEs. This is important for the numerical solution of the DDE because once the orders are high enough, the discontinuities will not interfere with the numerical method and we can stop tracking them.

In summary, our analysis of the simple linear first-order DDE (1.3) suggests that a long-time oscillatory solution is possible and that its initial history may have an effect on the short-time solution. Keeping these two properties in mind, we now introduce a variety of case studies appearing in diverse areas. The need for a *bifurcation diagram* where a property of the solution (extrema, period) is recorded as a function of a control parameter progressively appears as a priority.

1.2   Cyclic behaviors

A familiar example of delay-induced oscillations is when we try to adjust the shower temperature. The water flows at a uniform rate from the faucet to the shower head and we take this time to be τ seconds. We would never get into the shower before getting the temperature adjusted, but someone else might. Let $T(t)$ be the temperature at the faucet at time t and T_d is our desired temperature. We adjust the faucet based on the temperature at the faucet τ seconds ago and so the evolution of the temperature is described by

$$T' = -\kappa(T(t - \tau) - T_d).\tag{1.14}$$

The constant κ measures our reaction rate to a wrong temperature. A phlegmatic person would choose a small value of κ whereas an energetic person would prefer a large value of κ. But if κ is too small, the temperature will adjust very slowly and if κ is too large, oscillations may occur maybe with increasing amplitude leading to burns or frostbite. And how would we describe the "two-shower problem" when two showers share the same hot water resource?

Comparative studies between the observation of cyclic behaviors and the solution of a DDE are still rare. An example of such an analysis concentrates on the NFL football teams during the last 40 years. Banks [15] convincingly showed that the up and down of a football team experiences a simple periodicity that can be described by a DDE. See Figure 1.2. Banks considered the following linear DDE

$$\frac{dU}{dt} = b(\frac{1}{2} - U(t - \tau)),\tag{1.15}$$

where $0 < U < 1$ is defined as the decimal fraction of games won by an NFL team during one season. U is computed as follows. $U = (1 \times \#$ games won $+ \frac{1}{2} \times \#$ tied games $+ 0 \times \#$ games lost$)$/total number of games. $b > 0$

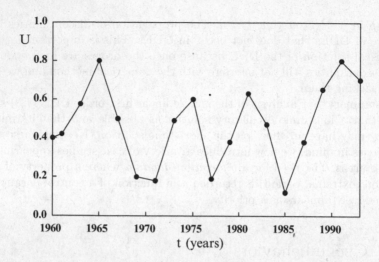

Figure 1.2: Performance of the Buffafo Bills, 1960 through 1992. Data taken from Banks [15]. Only the biennial values are shown. Separations between peaks are successively 10, 6, and 10 yr. Separations between troughs are successively 10, 6, and 8 yr. The average value of these six numbers gives a period close to 8 yr.

is defined as the growth rate. In this form, Eq. (1.15) says that the rate at which U changes at the present time is proportional to the difference between the average value $U = 1/2$ and the value of U at some previous time $t - \tau$. In other words, it takes a certain amount of time for a team to turn around for better or worse. What would be interesting to know is this "turnaround time" or delay τ. In his analysis, Banks found that the averaged periodicity between best performances of a team was close to 8 yr (see Figure 1.2.) Introducing $y \equiv U - 1/2$, we may rewrite Eq. (1.15) as Eq. (1.3) with $k = -b < 0$. Using then (1.7), we know that Eq. (1.15) admits the harmonic solution $U = 1/2 + A \sin(\omega t)$ if

$$\omega\tau = \pi/2 \quad \text{and} \quad b\tau = \frac{\pi}{2}. \tag{1.16}$$

Because the period P in Figure 1.2 is approximately 8 yr, we determine ω from the expression $P = 2\pi/\omega = 8$ yr. We find $\omega = \pi/4$ yr^{-1} and using (1.16), we obtain the delay τ as

$$\tau = 2 \text{ yr.} \tag{1.17}$$

There are, of course, many other factors that influence the success of a team (new coach, injuries of key players, player trades, etc.). But Eq. (1.15) is based on the idea that the growth of a human activity (here the performance of a football team) depends on its status at a previous time. If the performances are poor, we intend to invest in new resources to achieve a better result in the future. If, on the contrary, the performances are high, we are not likely to invest and we become more vulnerable with respect to competition. If the competition or the pressure to succeed is high, a cyclic behavior is inevitable. This idea has been discussed in a variety of businesses including the growth of scientific results. In this case, the delay represents, for example, the time needed to write an essential paper. Goffman and Harman [76] analyzed the list of publications in the field of symbolic logic and discovered an oscillatory pattern. They analyzed these observations as an epidemic process although a linear DDE such as Eq. (1.3) was later proposed [221].

1.3   Car-following models

It is hardly necessary to emphasize the importance of transportation in our lives. In the US, vehicles are driven an average of 10,000 miles per year for passenger cars and 50,000 miles per year for trucks on a highway system that comprises more than 4 million miles. The indices in other countries may be somewhat different, but the importance of the transportation system, and especially the highway component of it, is just the same or even greater. Traffic flow theories seek to describe in a precise mathematical way the interactions between the vehicles and their operators, and the highway system with all its operational elements. The scientific study of traffic flow had its beginnings in the 1930s with the application of probability theory to the description of road traffic but it was in the 1950s that major theoretical developments emerged using a variety of approaches, such as car-following, traffic wave theory (hydrodynamic analogy), and queueing theory [83, 33].

The following equation

$$x''_{n+1}(t + \tau) = \alpha(x'_n - x'_{n+1}) \qquad (1.18)$$

can be used for determining the location and speed of the following car (at $x = x_{n+1}$) given the speed pattern of the leading vehicle (at $x = x_n$). If a driver reacts too strongly (large value of α representing excessive braking) or too late (long reaction time τ), the spacing between vehicles may become unstable (i.e., we note damped oscillations in the spacing between vehicles). A typical solution of Eq. (1.18) for two cars is shown in Figure 1.3. The distance between the two cars dangerously drops from 10 m to about 5 m after the leading vehicle reduces its speed. If $\alpha\tau$ is increased, the number of damped oscillations increases. A sober driver needs about 1 s in order to start breaking in view of an obstacle. With 0.5 g/l alcohol in blood

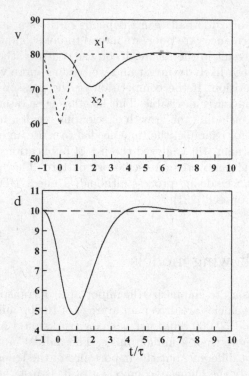

Figure 1.3: Car-following models. Top: Speed $v = x_1'$ and $v = x_2'$. Bottom: Distance $d = x_1 - x_2$ between the two cars. The lead vehicle reduces its speed of 80 km/h to 60 km/h and then accelerates back to its original speed with constant braking and acceleration rates. The initial spacing between vehicles is 10 m. $\alpha = 0.5$ s^{-1} and $\tau = 1$ s.

(2 glasses of wine), this reaction time is estimated to be about 1.5 s[3]. Figure 1.4 shows that the oscillations near the stable equilibrium increase.

It is clear that a driver's acceleration or deceleration also depends on the distance to the precedent car. The closer the driver is, the more likely the driver is to respond strongly to an observed relative velocity. The simplest way to model this is to let the sensitivity be inversely proportional to the distance. Many models considered the general function [33]

$$\lambda = c\frac{(x_n'(t + \tau))^m}{(x_{n-1} - x_n)^l}, \tag{1.19}$$

where c is a positive parameter and m, l are nonnegative parameters not necessarily integers. For example, $m = 0$ and $l = 1$ or 2.

---

[3]At a Blood Alcohol Content (BAC) of 0.8 g/l, (0.08 % BAC is the legal limit in the U.S. and in the U.K.), the average driver reaction time doubles from 1.5 to 3 s. In France, Belgium, and Germany the legal limit is 0.05 % BAC.

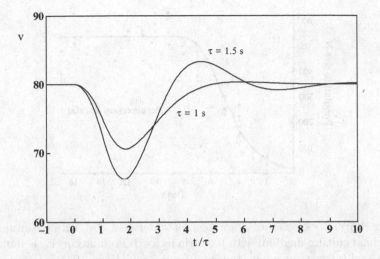

Figure 1.4: Alcohol and driving. Alcohol decreases the reaction time of driver 2 allowing more oscillations near the stable distance between driver 1 and driver 2.

1.4   Population dynamics

Pierre Francois Verhulst (1804–1849) was a Belgian professor of mathematics at the Université Libre de Bruxelles and at the Ecole Royale Militaire (also located in Brussels) in 1835. Forced by the administration to make a choice in 1840, he kept the more lucrative position at the Ecole Royale Militaire. Stimulated by Malthus's "Essay on the Principle of Population," Verhulst (1838) [237] published what he called a "logistique" equation[4] to describe the sigmoidal growth of population density to carrying capacity. See Mawhin [152] for an excellent review of the Verhulst legacy. The Verhulst equation was rediscovered by Pearl and Reed (1920) [184]. Then, Lotka (1925) [143] derived the same equation mathematically, calling it "the law of population growth," and the Russian biologist Gause (1934) [69] demonstrated its validity in laboratory experiments. See Kingsland [125] for a historical review. The so-called continuous logistic equation is given by

$$\frac{dN}{dt'} = rN(1 - \frac{N}{K}), \tag{1.20}$$

where r and K are defined as the growth rate and the carrying capacity of the population, respectively. The solution of Eq. (1.20) can be determined analytically because the equation is separable. It has a sigmoidal form starting exponentially from $N(0) \ll K$ and saturating at $N = K$.

---

[4]To presumably differentiate from the Malthusian "logarithmique."

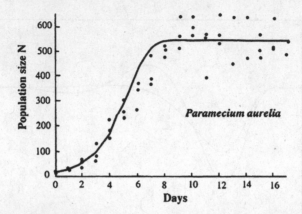

Figure 1.5: The growth of *Paramecium aurelia* in test tubes containing Osterhaut culture medium with bacteria as food. Population size is number per 0.5 ml (after Gause [69], redrawn from Case [45] p. 104).

The population growth of the protozoan *Paramecium* in test tubes is a typical example (Figure 1.5). Under the conditions of the experiment, the population stopped growing when there were about 552 individuals per 0.5 ml. The time points show some scatter, which is caused both by the difficulty in accurately measuring population size (only a subsample of the population is counted) and by environmental variations over time and between replicate test tubes. A linear regression of the data N'/N versus N gives $r = 0.99$ and $K = 552$.

The logistic equation (1.20) assumes that organisms' birth and/or death rates respond instantaneously to changes in population size. However, there are some organisms that exhibit pulses of reproduction and have some lag time (on the order of one generation) before they reproduce again. Delays occur if the organism stores the nutrient or due to the cell cycle or from environmental conditions (supply of food). Hutchinson [105] was one of the first mathematicians to introduce a delay into the logistic equation to account for hatching and maturation periods. He pointed out that the observed oscillations could be explained by a finite time delay in the crowding or resource term. Specifically, he studied the following equation

$$\frac{dN}{dt'} = rN(1 - \frac{N(t' - \tau)}{K}). \tag{1.21}$$

This equation can be rewritten in dimensionless form if we introduce

$$y \equiv N/K \quad \text{and} \quad t \equiv t'/\tau. \tag{1.22}$$

Eq. (1.21) then becomes

$$\frac{dy}{dt} = \lambda y(1 - y(t - 1)), \tag{1.23}$$

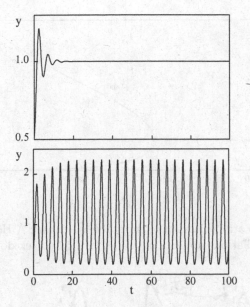

Figure 1.6: Oscillatory solutions of the logistic DDE. Top: $\lambda = 1$, the solution quickly decays to the stable steady state $y = 1$. Bottom: $\lambda = 1.8$, the oscillations grow and become sustained. The intial function is $y = 0.5$ ($-1 \leq s < 0$).

where

$$\lambda \equiv r\tau \qquad (1.24)$$

is the only parameter. Figure 1.6 represents the solution of Eq. (1.23) for two different values of λ. The figure suggests that the stable steady-state $y = 1$ becomes unstable at a value of λ between $\lambda = 1$ and $\lambda = 1.8$. We analyze the stability of $y = 1$ in Section 2.1 and find indeed a change of stability at

$$\lambda_c = \pi/2 \simeq 1.57. \qquad (1.25)$$

As $\lambda > \lambda_c$, the system transfers its stability to a stable period solution. We may represent the extrema of y as function of λ. See Figure 1.7. We note that the amplitude of the oscillations smoothly grows from zero. This is an example of a *bifurcation diagram* showing a *Hopf bifurcation* at $\lambda = \lambda_c$.

Lemming population cycles in the arctic north are nicely described by the logistic DDE with $r = 3.333/\text{yr}$ and $\tau = 9$ months ($\lambda = 3.333 \times 9/12 = 2.5$ is larger than λ_c). See Figure 1.8. When the time lag is 2.8 (or nearly three times r), the population overshoots K so much that it crashes down to extinction. This behavior is not unlike the population dynamics of reindeer introduced on two small islands in the Alaskan Pribiloff Islands (see Figure 1.9). Note, however, that the solution of the logistic DDE is still

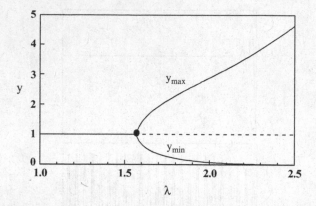

Figure 1.7: Bifurcation diagram of the stable solutions. A Hopf bifurcation to sustained oscillations appears at $\lambda = \lambda_c \simeq 1.57$ (black dot).

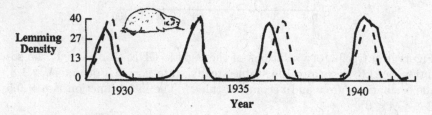

Figure 1.8: The full curve shows the density of lemmings in the Churchill area of northern Manitoba, Canada (number of individuals per hectare). The dashed curve is the solution of the logistic DDE with $r = 3.333$/year and $\tau = 0.72$ years (after May [154] redrawn from Case [45] p. 120).

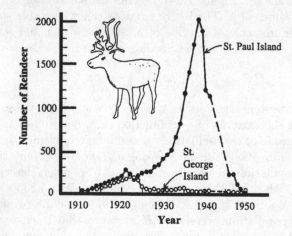

Figure 1.9: Introduced reindeer populations on two small islands in the Alaskan Pribiloff Islands (after Scheffer [208] redrawn from Case [45] p. 119).

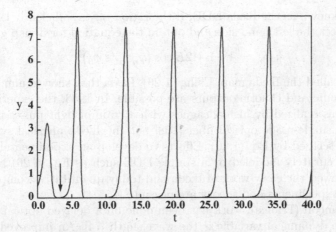

Figure 1.10: Strongly pulsating solution of the logistic DDE with $\lambda = 3$. The initial function is $y = 0.5$ $(-1 \leq s < 1)$. The first minimum is about 10^{-2} small (arrow) and the next minima are close to 10^{-6}. Under these conditions, the population physically disappears after the first pulse.

time-periodic if $\lambda \geq 3$ but the minima are so small that the value of y is meaningless in terms of population. See Figure 1.10. An asymptotic analysis of the oscillations for large values of λ is given by Fowler [66]. The incorporation of the delay in Eq. (1.21) allows us to describe the appearance of sustained oscillations in a single species population, without any predatory interaction of other species. However, the model description raises a number of questions such as the meaning of the finite time τ or why the delay enters the removal term y^2/K and not the production term y.

1.5 Nonlinear optics

In 1979, the Japanese physicist Kensuke Ikeda considered a nonlinear absorbing medium containing two-level atoms placed in a ring cavity and subject to a constant input of light. If the total length of the cavity is sufficiently large, the optical system undergoes a time-delayed feedback that destabilizes its steady-state output. Ikeda derived a set of coupled differential-difference equations from the Maxwell–Bloch equations [106]. Then introducing more assumptions, Ikeda formulated the following scalar DDE [107, 150]

$$\tau\phi' = -\phi + A^2\left[1 + 2B\cos\left(\phi(t - t_D) - \phi_0\right)\right] \tag{1.26}$$

which is known as the Ikeda DDE. If the ratio t_d/τ is sufficiently large, we may neglect the left hand side and obtain the equation for a map given by

$$\phi_n = A^2 \left[1 + 2B\cos\left(\phi_{n-1} - \phi_0\right)\right] \qquad (1.27)$$

which is called the Ikeda map. Using (1.26), Ikeda then showed numerically that periodic and chaotic outputs are possible. In 1983, the experimental system was realized by his colleagues with a train of light pulses injected in a long single-mode optical fiber [108] but the Ikeda physical system is poorly described by Eq. (1.26). Efforts to develop an experimental device that is accurately modeled by a simple DDE such as Eq. (1.26) immediately followed the early work of Ikeda and today quantitative comparisons between experiments and theory are possible.

In Besançon (France), work has been done on a delayed optical system where the dynamical variable is the wavelength [74]. An improved device using a tunable DBR laser was then realized [75, 138]. This experience then led to the development of a system based on coherence modulation. The dynamical variable is the optical-path difference in a coherent modulator driven electrically by a nonlinear delayed feedback loop [139]. The system is realized from an MZ coherence modulator powered by a short coherence source and driven by a nonlinear feedback loop that contains a second MZ interferometer and a delay line. In dimensionless variables, the response of the system is well described by [139],

$$\tau X' = -X + \beta \left[1 + \frac{1}{2}\cos(X(t - t_D) + \Phi)\right], \qquad (1.28)$$

where X is proportional to the optical-path difference. The bifurcation parameter β is proportional to the photodetector gain K which can be varied. The phase Φ can be changed electrically by means of a bias voltage applied to the first MZ. Experimentally, the ratio t_D/τ is chosen sufficiently large so that we may neglect the term in the left-hand side of Eq. (1.28). The resulting equation then is an equation for a map relating $X_{n+1} = X(t)$ and $X_n = X(t - t_D)$:

$$X_{n+1} = \beta \left[1 + \frac{1}{2}\cos(X_n + \Phi)\right]. \qquad (1.29)$$

The experimental bifurcation diagram is shown in Figure 1.11 for $\Phi = 0$. This bifurcation diagram is well reproduced by the numerical bifurcation diagram obtained from Eq. (1.29). Numerical and experimental values of the three marked points are compared in the next table.

	$\Phi = 0$		
Numerical	2.08	5.04	6.59
Experimental	2.07	5.30	6.69

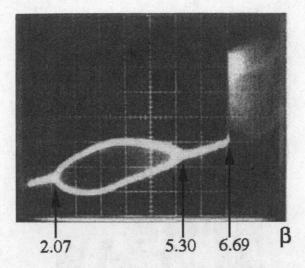

2.07   5.30   6.69   β

Figure 1.11: Experimental bifurcation diagram for $\Phi = 0$ from [139]. The points at $\beta = 2.07$ and $\beta = 5.30$ are two Hopf bifurcation points. The third point at $\beta = 6.69$ marks the transition to a "chaotic" output exhibiting erratic oscillations.

1.6   Fluid dynamics

DDEs appear in fluid mechanics when some memory effects need to be taken into account [238, 240]. We illustrate this by considering the case of vertical water fountains exhibiting oscillatory motion of their tips. Only for very low-momentum fluxes, ρu^2 ($\rho = 10^3$ kg m^{-3} is the fluid density and u is the initial velocity of the jet), the water exiting the fountains remains attached to the nozzle due to capillary and gravity forces. But, above a threshold of the momentum flux, a new regime is observed where the fluid detaches from the nozzle, forming an upward moving jet. The upward moving fluid then changes kinetic into potential energy until it reaches a maximum height (Figure 1.12 (1)) at which a lump of fluid begins to accumulate at the tip of the fountain. The maximum height that it can reach is given by $h = u^2/(2g)$ where g is the acceleration due to gravity. As the mass of the lump increases (Figure 1.12 (2)), the pull of gravity eventually overcomes the jet's momentum and the jet begins to fall. The backflow initially takes the form of a lump perched on the top of the jet and fed with liquid from below, falling under its own weight, flattening out the ascending column of fluid, and deforming under the inertial pressure of the jet (Figure 1.12 (3)). The lump eventually breaks up into a dispersed corolla (Figure 1.12 (4)), the jet re-emerges, and a new cycle begins (Figure 1.12 (5) to (8)). The overall effect is a pulsating motion of the jet, which is apparent not only to the eye but also to the ear (the breakup of the corolla

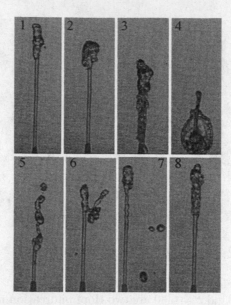

Figure 1.12: From left to right and from top to bottom: A period of the pulsation of a vertical fountain. $d = 3$ mm, $u_0 = 1.5$ ms^{-1}. The pictures are spaced by 2/25 seconds and the oscillation frequency is about 2 Hz. Note the corolla breakup of the gravity-induced backflow. Reprinted by permission from Macmillan Publishers Ltd (Nature) Villermaux [238] Copyright 1994.

is accompanied by a characteristic dripping sound). The pressure at the orifice of the jet is related to its instantaneous height and its variation oscillates with a dominant period. For fountains with moderate aspect ratios (height/diameter, $h/d \leq 50$), this oscillatory behavior develops before the onset of the Raleigh capillary breakup instability, which would otherwise cause the liquid column to fragment. If the fountain is oriented slightly away from the vertical, the backflow is no longer possible and the jet describes a parabola with a fixed maximum elevation. The gravity-induced backflow is thus essential for the onset of the oscillatory behavior. Villermaux [238] proposed that the oscillations are the result of the interplay between linear growth and a delayed nonlinear saturation and he mathematically models this mechanism by formulating a DDE for the amplitude $A(t)$ of the disturbances in the flow, equivalent to the fluctuating height of the fountain [239]

$$\frac{dA}{dt} = rA - \mu |A(t - \tau)|^2 A, \qquad (1.30)$$

where τ is the transit time through the recirculation loop. This timelag represents the interaction time of a fluid packet initially topping the fountain during its deformation until it breaksup in dispersed droplets and is estimated as the time for the packet to fall by a distance of the order of its own size (d) and independent of u_0 : $\tau \sim d/(dg)^{1/2} = (d/g)^{1/2}$. Eq. (1.30) rewritten in terms of $y \equiv \mu A^2/r$ and the new time $t' = t/\tau$ is equivalent to the delayed logistic equation (1.23) with $\lambda = 2r\tau$ and t' replacing t. We know that the solution is time-periodic if $\lambda > \pi/2$. Moreover, the frequency of the oscillations is independent of μ (because λ does not include μ after rescaling A^2) and approximate as $f \sim 1/(4\tau)$ if the amplitude of the oscillations is not too large. Villermaux [238] was criticized because he did not specify the parameter r and therefore no comparison can be made. The problem was later revived by Clanet [46] who proposed new experiments and a detailed physical model. He found that the frequency of the oscillations equals

$$f = \frac{g}{3u_0}. \tag{1.31}$$

Using the velocity $u_0 = 1.5 \text{ ms}^{-1}$ given by Villermaux [238], we find $f = 2.2$ Hz which compares well with the experimental observation of $f \sim 2$ Hz. It is thus possible to propose a complete physical model of the fountain oscillations. For other fluid dynamical phenomena such as turbulent flows, a detailed description is not always available and a global mathematical description using a DDE could be an interesting alternative [240].

1.7 Economics

The recurring fluctuations in economic activities (prices, output, inflation) that appeared in market economies since the spread of industrialization soon led to the idea that a business cycle is a self-sustained oscillation resulting from the lagged reaction of economic factors and the nonlinear relations between them. Kalecki's business cycle mode (1935) [118] is maybe the first mathematically detailed treatment of cyclical phenomena in economics. A key element of the theory is to posit a lag between the investment decision and installation of investment goods.[5] The Hopf bifurcation theory as a tool for demonstrating the emergence of a limit-cycle solution came to the attention of economists much later [233]. Today, DDEs have found their way in a variety of economical models [32] and bifurcation techniques are frequently used to analyze the effects of specific nonlinearities [19].

---

[5]Michal Kalecki's 1935 model used a linear difference-differential equation mix to yield cycles but his work in 1937 and 1939 used a nonlinear system to obtain limit-cycles. In economics, one should always publish in English. Although Kalecki claimed to have anticipated many of the principles stated in Keynes's *General Theory*, his articles (1933, 1935) were published in Polish and French and thus went unrecognized. Attempting to rectify this, Kalecki decided to publish a claim of precedence to Keynes in a 1936 article but in Polish again.

We illustrate the importance of delay in economy by first considering a problem that does not exhibit a cyclic behavior. We wish to describe the aggregate human capital stock $H(t)$ in terms of the individual's human capital. Overlapping generations offer new ideas and technologies and a long life working expectancy may not necessarily be interesting for advanced economies. But we need to take into account the time T devoted to schooling before entering the job market. To this end, de la Croix and Licandro [47] formulated the following DDE for $H(t)$

$$H' = \exp(-\beta T)TH(t - T) - \beta H. \qquad (1.32)$$

The last term on the right-hand side of Eq. (1.32) means that the aggregate human capital decreases at a rate β. This is compensated by the entry of new generations in the job market. At time t, $\exp(-\beta T)$ individuals of generation $t - T$ enter the job market with human capital $TH(t-T)$. The human capital is here assumed proportional to the time spent in school but also on $H(t)$ because we expect that the cultural ambiance of society at the time of birth will have a positive impact (through, e.g., the quality of the schools). Finally, the optimal time spent in school is not a constant but needs to take into account the effect of the loss of wage income if the entry on the job market is delayed and the negative effect of the instantaneous probability of death β. This is modeled by writing

$$T = \frac{1}{\beta_0 + \beta}. \qquad (1.33)$$

The important quantity to compute is the growth rate σ. The growth rate is determined by substituting $H = \exp(\sigma t)$ into Eq. (1.32). This then leads to an equation for σ called the characteristic equation (see Chapter 2). It is given by

$$\beta + \sigma = T \exp(-(\beta + \sigma)T). \qquad (1.34)$$

This equation admits the parametric solution

$$\beta = -\beta_0 + \sqrt{\exp(-x)/x}, \qquad (1.35)$$
$$\sigma = -\beta + (\beta_0 + \beta)x, \qquad (1.36)$$

and $\sigma = \sigma(\beta)$ is shown in Figure 1.13. Provided β_0 is sufficiently small, an increase of β from zero first leads to an increase of growth. After some point any increase of β implies a decrease of the growth. Simply saying, the effect of life expectancy on growth is positive for countries with a relatively low life expectancy, but could be negative in more advanced countries.

We next consider a problem that displays unusual oscillations. For the last 30 years, efforts have been directed to model foreign exchange rates. Several events have contributed to these research activities such as the adoption of the common currency Euro in 2002. A common feature of many models is to describe the exchange rate $S(t) = S_0(t) + x(t)$ as the sum of two

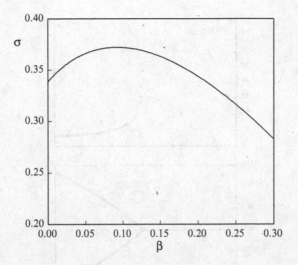

Figure 1.13: Growth rate of the aggregate human capital $H(t)$ as a function of the instantaneous probability of death β ($\beta_0 = 0.1$).

distinct quantities. $S_0(t)$ is defined as the "natural" foreign exchange rate based on pure economical factors. The second term $x(t)$ denotes the effect of perturbations that depend on noneconomical factors such as expectations, speculation, or random disturbances. The following DDE has been recently proposed by Erdélyi [56] as a minimal model describing the evolution of x,

$$x' = a\left[x - x(t-1) - |x|\,x\right]. \qquad (1.37)$$

The first two terms in Eq. (1.37) describe the growth of the exchange rate based on comparing rates at time t and at time $t-1$, respectively. If the exchange rate increases because $x(t) > x(t-1)$, it is worthwhile to purchase foreign currency. Hence, the demand for foreign currency goes up and the exchange rate will continue to increase. On the contrary, if the exchange rate decreases because $x(t) < x(t-1)$, the tendency will be to sell foreign currency and the demand will go down. At some time, a number of agents will realize that the absolute deviation $|x(t)|$ increases and will start to trade. The last term in Eq. (1.37) describes this effect. Because $x' = -x^2$ ($x > 0$) and $x' = x^2$ ($x < 0$), $|x|$ will continuously decrease. In reality, the depreciation (or appreciation) of the domestic current rate leading to a growth of $|x|$ and the rescuing nonlinear feedback are competing and we expect an oscillatory behavior. Figure 1.14 shows a bifurcation to a periodic solution that appears as soon as a passes 1. This bifurcation is not a Hopf bifurcation. As we approach a Hopf bifurcation point, the amplitude of the oscillations goes to zero and the period goes to a fixed quantity. Here, the amplitude of the oscillations goes to zero but the period

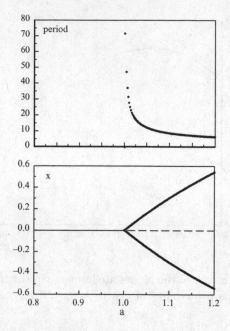

Figure 1.14: Bifurcation diagram of the periodic solutions emerging from $a = 1$. As $a \to 1^+$, the extrema of the oscillations decrease to zero but the period goes to infinity.

goes to infinity. By using asymptotic techniques, it is possible to show that the extrema of oscillations x_M and the period P scale such as $x_M \sim a - 1$ and $P \sim (a-1)^{-1/2}$, respectively, as $a \to 1^+$ [62].

1.8   Mechanical engineering

Gantry cranes can lift several hundred tons and can have spans of well over 50 meters. See Figure 1.15. For fabrication and freight-transfer applications, it is important for the crane to move payloads rapidly and smoothly. If the gantry moves too fast the payload may start to sway, and it is possible for the crane operator to lose control of the payload. At the 2005 ASME meeting, the question was raised whether a delayed feedback control for container cranes could be more efficient than conventional techniques. Henry et al [90] and Masoud et al [156, 157, 158] developed a control strategy based on a time-delayed position feedback of the payload cable angles. Its goal was to significantly reduce the sway at the end of motion.

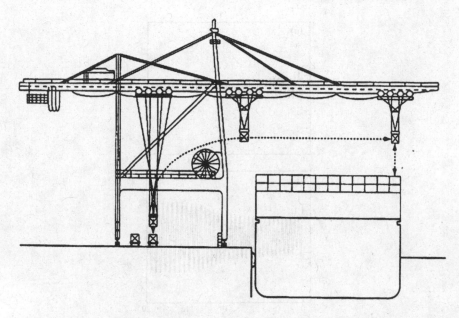

Figure 1.15: Container crane and ship (from H. Park and K.-S. Hong [181]).

The formulation of the pendulum model for the container crane is described in Section 2.3. In its simplest form, it is given by

$$y'' + \varepsilon y' + \sin(y) = -k\cos(y)(y(t-\tau) - y), \qquad (1.38)$$

where y represents the angle. The left-hand side models a weakly damped oscillator and the right-hand side is the contribution of the feedback control. Without going into details, it is not too difficult to understand why a delayed feedback may have a stabilizing effect. Near the equilibrium solution $y = 0$, $\sin(y) \sim y$ and $\cos(y) \sim 1$ and Eq. (1.38) can be rewritten as

$$y'' + \varepsilon y' + y = -k(y(t-\tau) - y). \qquad (1.39)$$

If now τ is small, we expand the delayed variable as $y(t-\tau) \sim y - \tau y'$ and reformulate Eq. (1.39) as

$$y'' + (\varepsilon - k\tau)y' + y = 0. \qquad (1.40)$$

The effect of the delay appears in the damping coefficient. If $k < 0$, this coefficient increases with the delay allowing faster decaying oscillations. This conclusion is valid provided y is sufficiently close to zero. On the other hand, if the size of y is arbitrary, then we need to worry about nonlinear effects. Figure 1.16 illustrates the possible danger of using a delayed feedback control. For small amplitude perturbation of the equilibrium position $y = 0$, the oscillations are damped. On the other hand, if the perturbation is large enough, the oscillations of the pendulum become sustained.

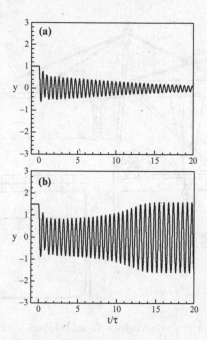

Figure 1.16: The values of the fixed parameters are $\tau = 12$, $\varepsilon = 0.1$, and $k = -0.15$. (a): $y'(0) = 0$ and $y = 1$ $(-\tau < t < 0)$; (b): $y'(0) = 0$ and $y = 1.5$ $(-\tau < t < 0)$.

1.9 Combustion engines

Improving the control of the air-to-fuel ratio (A/F) allows gasoline port-fuel injection engines to meet more stringent emission regulations. With the growing use of Universal Exhaust Gas Oxygen (UEGO) sensors more flexible air-to-fuel ratio control architectures capable of achieving low emission levels can be implemented. The delay between the fuel injection and UEGO sensor measurement can, however, limit the performance of the air-to-fuel ratio control loop. Figure 1.17 (right) shows a plot of measured engine air-to-fuel ratio in response to a step in the fuel injection rate from an engine operating at 1000 rev/min (rpm) and at constant air flow.

The two relevant quantities are $F_{ac}(t)$ and $F_{fc}(t)$ defined as the airflow rate and the fuel flow rate into the engine cylinders, respectively. The air-to-fuel ratio is then defined as $R(t) \equiv F_{ac}/F_{fc}$ and the control objective is to regulate R to the desired value R_d. The amount of fuel available to the engine is determined by the injection fuel flow rate $F_{fi}(t)$. Averina et al. [10] considered a simple mathematical model for the A/F control problem that takes into account the signal coming from the UEGO sensor. The latter is described in terms of $x_{sen} = R_{sen}^{-1}$ and is related to $x = R^{-1}$ via first-order dynamics

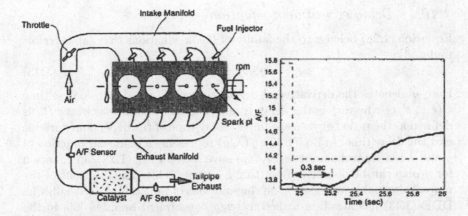

Figure 1.17: Left: Gasoline engine. Right: Delay from fuel injection step (broken line) to A/F changes at 1000 rpm. The delay varies with engine speed, mass flow rate through the engine, exhaust pressure, and exhaust temperature. At 2000 rpm, the delay reduces to 0.2 s and to 0.1 s at 3000 rpm (redrawn from Averina et al. [10], copyright 2005 IEEE).

$$\frac{dx_{sen}}{dt} = -a(x_{sen} - x(t - t_d)), \tag{1.41}$$

where a (~ 5 s^{-1})and t_d (~ 0.25 s) correspond to the rate and the delay of the A/F sensor. Assuming a constant airflow rate ($dF_{ac}/dt = 0$), the evolution of the deviation $y = x_{sen} - x_d$, where $x_d = R_d^{-1}$ is given by

$$y'' + y'(a + \tau^{-1}) + a\tau^{-1}y = ky(t - t_d) \tag{1.42}$$

where prime means differentiation with respect to time t, τ (~ 0.2 s) is the fuel evaporation rate from the liquid puddle, and $k < 0$ is the gain of the controller command.

1.10   Classes of DDEs

In the previous examples, we saw two different bifurcations to a periodic solution, we noted that square-wave oscillations are possible if the delay is large, and we also learned how delay could have a stabilizing effect. Mathematicians have tried to classify DDEs by their difficulty in order to identify some of their key properties. The problems that are the most often discussed are first-order nonlinear DDEs exhibiting square-wave oscillations and second-order nearly conservative equations exhibiting both periodic and quasiperiodic oscillations.

1.10.1 Delay recruitment equation

Equation (1.28) belongs to the family of Ikeda equations that can be reformulated as

$$\varepsilon y' = -y + f(\lambda, y(t-1)). \tag{1.43}$$

Here y' denotes the derivative of y with respect to the dimensionless time t ($t \equiv t'/t_d$ where t_d is the delay time). The fixed parameter $\varepsilon \equiv \tau/t_d$ is defined as the ratio between the linear decay time of the dependent variable and the delay time. In Eq. (1.43), $f(\lambda, y)$ represents a nonlinear function of y and λ is a control parameter. As we have seen for Eq. (1.28), an equation for a map can be obtained by setting $\varepsilon = 0$ in (1.43). The solutions of the map give valuable information on the solutions of Eq. (1.43) for ε small. The DDE (1.43) has also been called a *delay recruitment equation* [65] in the context of biological or medical applications. Mackey [147] (see Chapter 3) formulated an equation of the form (1.43) where

$$f(\lambda, y) = \frac{\lambda}{1 + y^p} \tag{1.44}$$

describes a negative feedback ($p > 0$). Note that Eq. (1.43) exhibits a simple damping term ($-y$) and that the nonlinear function f only depends on $y(t-1)$. The linear DDE $\varepsilon y' = \pm y(t-1)$ and the delayed logistic equation $\varepsilon y' = \lambda y(1 - y(t - \tau))$ do not belong to this class of equations. The bifurcation diagram of the Ikeda equation with

$$f(\lambda, y) \equiv \lambda(1 - \sin(y)) \tag{1.45}$$

has been studied in detail and has revealed a large number of bifurcation transitions [151, 173]. The parameter ε is naturally small and is responsible for a sharp Hopf bifurcation transition [61] (see Figure 1.18). Other functions $f(\lambda, y)$ have been investigated. Shanz and Pelster [207] studied Eq. (1.43) with

$$f(\lambda, y) = -\lambda \sin(y) \tag{1.46}$$

where $\lambda > 0$ and y is defined as a phase variable (first-order phase locked loop describing the synchronization of two oscillators (see Chapter 8). Hong et al. [99] investigated Eq. (1.43) with

$$f(\lambda, y) = 1 - \lambda y^2. \tag{1.47}$$

1.10.2 Harmonic oscillator with delay

Another class of DDEs concentrates on the harmonic oscillator with delayed forcing [40] and has the form

$$y'' + ax' + y = f(y(t - \tau), y'(t - \tau)), \tag{1.48}$$

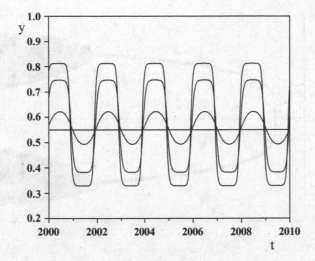

Figure 1.18: Periodic solutions of Ikeda DDE near the first Hopf bifurcation point. $\varepsilon = 10^{-2}$ and, from small to large amplitude, $\lambda = 1.17,\ 1.18,\ 1.19,\ 1.20$. The Hopf bifurcation point is located at $\lambda_H = 1.177$. Note the rapid change of the oscillations from harmonic to square-wave as the deviation $\lambda - \lambda_H = O(\varepsilon)$ progressively increases from zero.

where the function $f(y, y')$ is nonlinear. In a mechanical system subject to a delayed feedback, y and y' represent the position and velocity at time t. If the function f only depends on $y(t - \tau)$, we talk about *position feedback* whereas if f only depends on y', we refer to the case of *velocity feedback*. Machine tool vibrations have negative effects on the quality of machined surfaces. One of the most important causes of undesired vibrations in the cutting process is the so-called regenerative effect. Its physical basis is a time delay that arises naturally in the cutting process, where the delay is inversely proportional to the cutting speed (see Chapter 6). Johnson and Moon [115] investigate an electromechanical system and simulate their experiments by studying the following equation,

$$y'' + ay' + b(y - y^3) = c(y' - y'(t - 1)). \qquad (1.49)$$

They found periodic, quasiperiodic, and chaotic oscillations. The values of the parameters were $a = 2.623$, $b = 170\pi^2$, and c is the control parameter. Figure 1.19 shows the bifurcation diagram where a Hopf bifurcation is followed by a bifurcation to quasiperiodic oscillations. Quasiperiodic oscillations are oscillations characterized by two noncommensurable frequencies. As a consequence the maxima (minima) are changing at each oscillation but are bounded by an upper and lower limit.

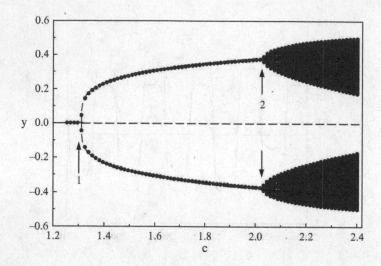

Figure 1.19: Bifurcation diagram of the stable solutions of Johnson and Moon equations. Points 1 and 2 mark a Hopf bifurcation followed by a bifurcation to quasiperiodic oscillations.

The so-called sunflower equation

$$y'' + \frac{a}{\tau}y' + \frac{b}{\tau}\sin(y(t-\tau)) = 0 \qquad (1.50)$$

was originally proposed by Israelsson and Johnsson [113] to describe the geotropic circumnutations of *Helianthus annus*. More than a century ago, plant physiologists were aware that elongating plant organs–roots, shoots, branches, flower stalks – rarely grow in just one direction. The organ's instantaneous growth direction usually oscillates slowly about a mean growth direction. The plant organ tip, as seen from a distant viewpoint, describes an ellipse and, in three dimensions, the tip traces a helix. Such "circumnutations" of sunflower seedlings were modelled in 1967 by Israelsson and Johnsson [113]. According to the model, the movement was completely dependent on gravity. However, a Spacelab experiment in 1983 showed that under microgravity conditions oscillations were still occurring. They are, however, less regular [37, 116]. Here, y denotes the angle with the plumb line. Using numerical simulations, Johnsson [117] determined the values of a and b for which Eq. (1.50) has a numerical periodic solution with period 157 minutes assuming a delay τ of 30 minutes. The equation was later studied mathematically by Somolinos [216] who showed that for $a = 4.8$ and $b = 0.186$, there is a periodic solution for τ between 35 and 80 minutes.

1.11   Analytical tools

Many of the problems that we are facing with DDEs involve such difficulties as transcendental equations or nonlinear evolution equations that preclude solving them exactly. Consequently, solutions are approximated using numerical techniques, analytic techniques, and combinations of both. Foremost among the analytic techniques are the systematic methods of perturbations (asymptotic expansions) in terms of a small or a large parameter. In this book, several such techniques are highlighted. They have been developed for solving particular DDE problems and they are listed in the following table.

Methods	Sect.
Linear stability analyses	2
Construction technique for a single Hopf bifurcation	3.2, 6.6.2
Piecewise constant nonlinearities	2.6, 3.3
Weakly perturbed strongly nonlinear relaxation oscillators	5.3.2
Multiple time-scale methods for weakly nonlinear oscillators	6.2, 6.3
Weakly perturbed strongly nonlinear conservative oscillators	7.2.1
Construction technique for a double Hopf bifurcation	7.2.3

One can't do everything, however. If the model equations are too complex, numerical approaches are needed. But it is important to make clear that asymptotic methods are available, analytical tools capable of extracting information of physical significance.

2

Stability

The industrial revolution in Europe followed the introduction of prime movers, or self-driven machines. It was marked by the invention of advanced grain mills, furnaces, boilers, and the steam engine. These devices could not be adequately regulated by hand, and so arose a new requirement for automatic control systems. A variety of control devices was invented, including float regulators, temperature regulators, pressure regulators, and speed control devices. In the mid-1800s mathematics was first used to analyze the stability of these feedback control systems. In 1840, G.B. Airy discovered that an improper design of the feedback control loop leads to wild oscillations. He was the first to discuss this instability of the control system by using differential equations [1]. Later, J.C. Maxwell in 1868 analyzed the stability of steam engine regulating devices then known as governors [159]. His technique was to linearize the differential equations of motion to find the characteristic equation of the system. He studied the effect of the system parameters on stability and showed that the system is stable if the roots of the characteristic equation have negative real parts.[1]

---

[1]Maxwell raised the mathematical question of whether a given polynomial of order n and real coefficients has roots with negative real parts and if we could find a solution that can be expressed solely in terms of the coefficients, thus avoiding the explicit computations of the roots. He was not aware that the problem had already been solved in 1856 by Hermite [91]. In 1877, the applied mathematician E. J. Routh provided a numerical technique for determining when a characteristic equation has stable roots [201]. Unaware of the work of Maxwell and Routh, A. B. Stodola posed the problem of determining the stability of the characteristic equation to A. Hurwitz [104] in 1895. Hurwitz

T. Erneux, *Applied Delay Differential Equations*, Surveys and Tutorials in the Applied Mathematical Sciences 3, DOI 10.1007/978-0-387-74372-1_2, © Springer Science+Business Media, LLC 2009

The principal difficulty in studying DDEs lies in the transcendental character of the characteristic equation leading to an infinite number of complex roots. A delay problem connected to the position control of mechanical devices where the number of roots is finite is analyzed below. But, in general, we need to solve the characteristic equation using numerical methods and graphical tools. Often, we are interested in studying the bifurcation diagram of the pulsating solutions in a finite domain of parameters. Then, most of the difficulties of determining the complete spectrum can be set aside, because only a few eigenvalues will contribute to the observed oscillations. In the next section, we analyze the characteristic equation of a simple linear DDE and identify particular points where a change of stability occurs.

2.1   The characteristic equation

We wish to determine all the solutions of a linear DDE such as Eq. (1.3). Redefining the time variable as $t \rightarrow t/\tau$, Eq. (1.3) can be rewritten in a simpler form as

$$\frac{dy}{dt} = ay(t-1), \tag{2.1}$$

where

$$a \equiv k\tau \tag{2.2}$$

is our control parameter. Eq. (2.1) is linear which suggests trying an exponential solution of the form

$$y = c \exp(\sigma t). \tag{2.3}$$

Substituting (2.3) into Eq. (1.3) leads to an equation for the growth rate σ, called the *characteristic equation*, given by

$$\sigma - a \exp(-\sigma) = 0. \tag{2.4}$$

Equation (2.4) is a *transcendental equation* and admits several roots.[2] We separate the case σ real and the case σ complex.

---

gave a solution in terms of determinants on the basis of the Hermite paper. Modern proofs may be found in Uspenky [236].

    [2]The solution of this equation is known in terms of the Lambert function $W(x)$ that satisfies the equation $W(x)\exp(W(x)) = x$. The solution of Eq. (2.4) with a real then is $\sigma = W(a)$. In symbolic software packages such as Maple and MATLAB, $W(x)$ is a standard function now.

2.1.1 Roots

1. σ is real. From Eq. (2.4), we have the implicit solution

$$a = \sigma \exp(\sigma). \qquad (2.5)$$

Studying the function $a = a(\sigma)$ given by (2.5), we find that σ is a single positive root if $a > 0$ and that there exist two distinct negative roots if $a_c < a < 0$, where $a_c \equiv -e^{-1}$. If $a = a_c$, we have a double root ($\sigma = -1$) and, if $a < a_c$, there exist no real roots. See Figure 2.1.

2. σ is complex. Substituting $\sigma = \sigma_r + i\sigma_i$ into Eq. (2.4) and separating real and imaginary parts, we obtain two equations for σ_r and σ_i given by

$$\sigma_r - a \exp(-\sigma_r) \cos(\sigma_i) = 0, \qquad (2.6)$$

$$\sigma_i + a \exp(-\sigma_r) \sin(\sigma_i) = 0. \qquad (2.7)$$

Eliminating the common coefficient $a \exp(-\sigma_r)$ leads to the following equation

$$\cot(\sigma_i) = -\frac{\sigma_r}{\sigma_i} \qquad (2.8)$$

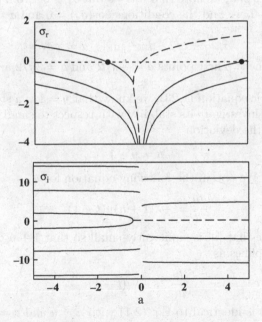

Figure 2.1: The solutions of the characteristic equation. The full and broken lines correspond to σ complex and σ real, respectively. Each σ exhibits the limit $\sigma_r \to -\infty$ as $|a| \to 0$. We note that all the σ_r are negative in the interval $-\pi/2 \le a \le 0$ meaning stability of the zero solution. The two dots mark the point where one σ_r changes sign.

that contains no parameter. Using (2.8) and then (2.7), the solution can be analyzed in parametric form as (σ_i is the parameter)

$$\sigma_r = -\sigma_i \cot(\sigma_i), \tag{2.9}$$

$$a = -\frac{\sigma_i \exp(\sigma_r)}{\sin(\sigma_i)}. \tag{2.10}$$

See Figure 2.1.

In summary, the solution of Eq. (2.1) can be described as a sum of exponentials of the form

$$y = \sum_n c_n \exp(\sigma_n t) \tag{2.11}$$

where the c_n are unknown. The coefficients c_n can be determined in terms of the initial function $y_0(t)$ ($-1 \leq t < 0$) using the Laplace transform [21]. Practically, we wish to know if $y \to 0$ as $t \to \infty$ meaning that the determination of the σ_n is good enough ($Re(\sigma_n) < 0$ for all n).

2.1.2 Hopf bifurcation point

At critical values of a, we note that $\sigma_r = 0$ but $\sigma_i \neq 0$. From Eqs. (2.6) and (2.7) with $\sigma_r = 0$, we find the conditions $\cos(\sigma_i) = 0$ and $a = -\sigma_i/\sin(\sigma_i)$ which imply

$$\sigma_i = \pm\pi/2 + k\pi \quad \text{and} \quad a = \mp\sigma_i \tag{2.12}$$

where $k \in Z$. The two first points $a = -\pi/2$ and $a = 3\pi/2$ are indicated in Figure 2.1.

For the logistic equation (1.23), we know that $y = 1$ is a steady-state solution. We may investigate its stability with respect to small perturbations by introducing the deviation

$$u = y - 1 \tag{2.13}$$

into Eq. (1.23). We obtain the following equation for u,

$$\frac{du}{dt} = -\lambda(1 + u)u(t - 1). \tag{2.14}$$

We next assume that $|u|$ is sufficiently small so that $1 + u \simeq 1$. Equation (2.14) then simplifies as

$$\frac{du}{dt} = -\lambda u(t - 1). \tag{2.15}$$

Equation (2.15) is identical to Eq. (2.1) with $y = u$ and $a = -\lambda$. Because the zero solution is stable in the interval $-\pi/2 < a < 0$, we conclude that $y = 1$ is stable if

$$0 < \lambda < \pi/2. \tag{2.16}$$

The critical point $\lambda = \pi/2$ is a Hopf bifurcation point that leads to a branch of periodic solutions (see Chapter 3).

2.2   Position control and sampling

Position control is a frequent mechanical controlling problem in robotics. The aim is to drive the robot arm into a desired position. To achieve a clear picture about the behavior of the control, digital effects, such as sampling, should also be included in the mechanical model. Sampling is a kind of delay in information transmission that often leads to unstable oscillations. Analytical investigations of simple models with one degree of freedom play a central role in understanding technical phenomena and designing a safe system. Another example of sampling is described in Chapter 5. Here, we reproduce the analysis by Insperger and Stépán [110] of a simple position control problem. Only the characteristic equation is modified in order to use Hurwitz stability conditions.

Because of the digital sampling effect, the evolution equation is a DDE but the stability problem can be reduced to a finite eigenvalue problem. The system is described by (see Figure 2.2)

$$M\frac{d^2x}{dt'^2} = Q, \tag{2.17}$$

where prime means differentiation with respect to time t'. The sampling time is τ. At each time $t' = n\tau$, the control force Q is quasi-instantaneously readjusted in terms of the observed position $x(t_n)$ and observed velocity $dx(t_n)/dt'$. The control law is

$$Q = -Px(t_n) - D\frac{dx}{dt'}(t_n), \tag{2.18}$$

where P and D are positive coefficients. Introducing the dimensionless time

$$t = t'/\tau, \tag{2.19}$$

Eqs. (2.17) and (2.18) take the simpler form

$$x'' = -px(n) - dx'(n), \tag{2.20}$$

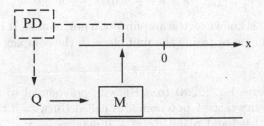

Figure 2.2: Position control. The position of mass M is sensed and a control force Q is applied to push the mass into the desired position. The control PD has proportional and differential gains (redrawn from [110]).

where prime means differentiation with respect to time t. The dimensionless parameters p and d are defined by

$$p = \frac{P\tau^2}{M} \quad \text{and} \quad d = \frac{D\tau}{M}. \tag{2.21}$$

Knowing position $x_n = x(n)$, velocity $v_n = x'(n)$, and acceleration $a_n = x''(n) \equiv -px(n) - dx'(n)$ at time $t = n$, we integrate Eq. (2.20) and obtain

$$x'' = a_n, \tag{2.22}$$

$$x' = v_n + a_n(s - n), \tag{2.23}$$

$$x = x_n + v_n(s - n) + \frac{a_n}{2}(s - n)^2. \tag{2.24}$$

Consequently, we determine $x_{n+1} = x(n+1)$, $v_{n+1} = x'(n+1)$, and $a_{n+1} = x''(n+1)$ at time $t = n+1$. The resulting equations form a system of three first-order difference equations of the form

$$\begin{pmatrix} x_{n+1} \\ v_{n+1} \\ a_{n+1} \end{pmatrix} = \begin{pmatrix} 1 & 1 & \frac{1}{2} \\ 0 & 1 & 1 \\ -p & -d & 0 \end{pmatrix} \begin{pmatrix} x_n \\ v_n \\ a_n \end{pmatrix}. \tag{2.25}$$

We wish to analyze the stability of the zero solution. To this end, we seek a solution of the form

$$x_{n+1} = zx_n, \quad v_{n+1} = zv_n \quad \text{and} \quad a_{n+1} = za_n, \tag{2.26}$$

where z is called the amplification factor. Substituting (2.26) into (2.25), we obtain the following homogeneous system of equations for x_n, v_n and a_n,

$$\begin{pmatrix} 1-z & 1 & \frac{1}{2} \\ 0 & 1-z & 1 \\ -p & -d & -z \end{pmatrix} \begin{pmatrix} x_n \\ v_n \\ a_n \end{pmatrix} = 0. \tag{2.27}$$

This system of equations has a nontrivial solution if the determinant of the coefficients vanishes. Expanding the determinant as a polynomial in z yields

$$z^3 - 2z^2 + z(1 + \frac{p}{2} + d) + \frac{p}{2} - d = 0. \tag{2.28}$$

This polynomial is known as the amplification polynomial. It is called stable if all the roots lie on or inside the unit circle in the complex z plane:

$$|z_1| \leq 1, \quad |z_2| \leq 1, \quad |z_3| < 1. \tag{2.29}$$

We next transform Eq. (2.28) to a Hurwitz polynomial to apply a more traditional stability test. A polynomial is called Hurwitz if the location of its roots in the left-hand plane $Re(s) \leq 0$ determines stability. To transform Eq. (2.28) to a Hurwitz polynomial, we use the conformal involutory transformation

$$z = \frac{1+s}{1-s}. \tag{2.30}$$

Inserting (2.30) into (2.28), we obtain

$$s^3(4+2d) + s^2(4+p-4d) + s(-2p+2d) + p = 0. \tag{2.31}$$

The Routh—Hurwitz stability conditions for the third-order polynomial

$$b_3 s^3 + b_2 s^2 + b_1 s + b_0 = 0 \tag{2.32}$$

are given by

$$b_1 > 0, \quad b_1 b_2 - b_0 b_3 > 0 \quad \text{and} \quad b_3 > 0. \tag{2.33}$$

The last condition is always satisfied because $d > 0$. The first condition requires that $d > p$ and the second condition leads to the inequality

$$p^2 + p(6 - 4d) + 4d(d - 1) < 0, \tag{2.34}$$

or equivalently,

$$0 \le p < p_H = -3 + 2d + \sqrt{9 - 8d}$$

because $p \ge 0$. The critical point $p = p_H$ corresponds to a Hopf bifurcation. This can be verified by substituting $s = i\omega$ ($\omega \ne 0$) into Eq. (2.32) and separating the real and imaginary parts. We find the two conditions $-b_3\omega^2 + b_1 = 0$ and $-b_2\omega^2 + b_0 = 0$, or equivalently,

$$b_1 b_2 - b_0 b_3 = 0 \quad \text{and} \quad \omega^2 = b_1/b_3 > 0. \tag{2.35}$$

The first condition is verified by $p = p_H(d)$ and the second condition provides the square of the Hopf bifurcation frequency

$$\omega^2 = \frac{p_H}{1 - 2d + \sqrt{9 - 8d}} > 0. \tag{2.36}$$

See Figure 2.3.

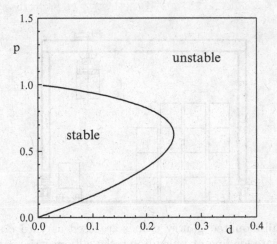

Figure 2.3: Position control. The Hopf bifurcation line $p = p_H$ delimits the region of stability.

2.3   Reduction of payload oscillations

Gantry cranes are used for moving objects within shipyards, ports, railyards, factories, and warehouses. See Figure 2.4. Gantry cranes can lift several hundred tons and can have spans of well over 50 meters. For fabrication and freight-transfer applications, it is important for the crane to move payloads rapidly and smoothly. If the gantry moves too fast the payload may start to sway, and it is possible for the crane operator to lose control of the payload. During the last four decades, different strategies of controlling payload pendulations without including the operator in the control loop have been investigated. Recently, the question was raised whether a delayed feedback control could be superior to conventional techniques. Henry et al [90] and Masoud et al [156]–[158] developed a control strategy based on a time-delayed position feedback of the payload cable angles. The efficiency of this technique was investigated by both numerical simulations of detailed mathematical models and by experiments in the laboratory [90, 174]. In [60], we have analyzed the possible bifurcations of the crane–payload system subject to a delayed control. We have shown that because of subcritical bifurcations, stable time-periodic attractors may coexist with a stable equilibrium. For safe control of the crane pendulations, such time-periodic regimes should be avoided by using a physical model and finding conditions for safe operation.

In [60], the model was derived using a Lagrangian approach. In this section, we consider a simpler model using Newtonian laws. The main forces controlling the crane–payload system are displayed in Figure 2.5. We assume that the cable is inextensible or its length is slowly varying

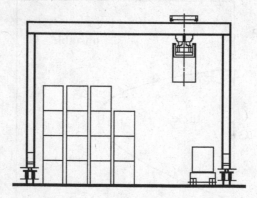

Figure 2.4: Rail-mounted gantry cranes are used as yard cranes. Except for the immediate loading or unloading tasks, all operations must be done automatically with an efficient anti-sway control technique (from Erneux and Kalmár-Nagy [60]).

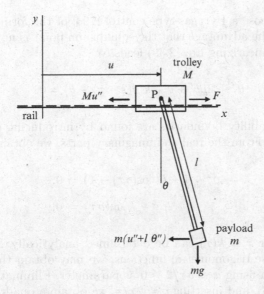

Figure 2.5: Simplest pendulum model for a container crane.

compared to the time-scale of the payload oscillations. The crane is assumed to ride on frictionless rails and the payload is assumed to rotate about a frictionless pivot P. The force applied to the motor is $F(t)$. The inertial force on the crane is Mu''. The inertial force on the payload, in the horizontal direction, is $m(u'' + l\theta'')$ and the gravity force on the payload is simply mg. Balancing forces in the horizontal direction ($\Sigma F_x = 0$) gives

$$Mu'' + m(u'' + l\theta'') = F \qquad (2.37)$$

and balancing the moments about the pivot point P of the payload ($\Sigma M_P = 0$) leads to

$$m(u'' + l\theta'')l\cos(\theta) + mgl\sin(\theta) = 0. \qquad (2.38)$$

Using (2.37), we eliminate u in Eq. (2.38) and obtain

$$\theta'' + \tan(\theta) + h(s) = 0, \qquad (2.39)$$

where prime means differentiation with respect to the dimensionless time $s \equiv \omega t$ and ω is the crane–payload frequency defined by $\omega \equiv \sqrt{(M+m)g/(Ml)}$. The external force is $h(s) \equiv F(s)/((M+m)g)$. Finally, we introduce a small damping term $(2\mu\theta')$ to take into account weak dissipation. Equation (2.39) then becomes

$$\theta'' + \tan(\theta) + 2\mu\theta' + h(s) = 0. \qquad (2.40)$$

We next propose a Pyrygas-type control [194] of the form $h = k(\theta(s - \tau) - \theta)$. It has the advantage that the equilibrium point is not modified by the feedback. Linearizing Eq. (2.39) leads to

$$\theta'' + 2\mu\theta' + \theta + k(\theta(s - \tau) - \theta) = 0. \qquad (2.41)$$

The linear stability boundaries are found by introducing $\theta = \exp(i\sigma s)$ into Eq. (2.41). From the real and imaginary parts, we obtain

$$-\sigma^2 + 1 + k(\cos(\sigma\tau) - 1) = 0, \qquad (2.42)$$

$$2\mu\sigma - k\sin(\sigma\tau) = 0. \qquad (2.43)$$

The solution for $k = k(\tau)$ can be determined analytically. If we wish to avoid the inverse trigonometric functions, we may obtain the solution in parametric form using $x \equiv \sigma\tau/2 \geq 0$ as parameter. Eliminating k in Eqs. (2.42) and (2.43), and inserting $\sigma = 2x/\tau$, we obtain a quadratic equation for τ. It always admits a positive real root given by

$$\tau = 2\left[\mu x\tan(x) + |x|\sqrt{\mu^2\tan^2(x) + 1}\right]. \qquad (2.44)$$

Having $\tau(x)$, we determine k using (2.43) with $\sigma = 2x\tau^{-1}$:

$$k = \frac{4\mu x}{\tau\sin(2x)}. \qquad (2.45)$$

By continuously increasing x from zero, the successive Hopf bifurcation curves are generated by (2.44) and (2.45) (full lines in Figure 2.6). The friction coefficient μ is generally small and if $\mu = 0$, the expressions of the Hopf bifurcation lines considerably simplify. From Eqs. (2.42) and (2.43), we find the following three cases

$$k_0 = 0 \text{ and } \sigma_0 = 1, \qquad (2.46)$$

$$\tau_0 = 2n\pi \text{ and } \sigma_0 = 1, \qquad (2.47)$$

$$k_0 = \frac{1}{2}\left[1 - (\frac{(2n+1)\pi}{\tau})^2\right] \text{ and } \sigma_0 = \frac{(2n+1)\pi}{\tau}. \qquad (2.48)$$

where $n = 0, 1, 2, \ldots.$ The horizontal line $k = 0$, the vertical line defined by (2.47) with $n = 1$, and the lines defined by (2.48) with $n = 0$ and 1 are shown by broken lines in Figure 2.6.

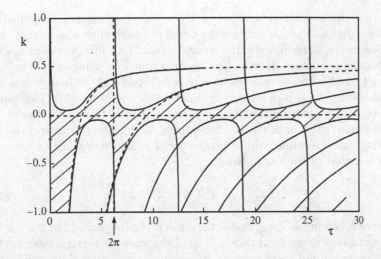

Figure 2.6: Successive Hopf bifurcation lines (solid) in the k versus τ parameter plane for $\mu = 0.025$. The broken lines correspond to the limiting case of no friction ($\mu = 0$) and are shown for the first four Hopf bifurcations. The crosshatched domain corresponds to a stable steady state.

2.4 Traffic stability

The control of traffic congestion problems is an important problem in our society: where to install traffic lights or stop signs, how many lanes to build for a new highway, should we develop alternate forms of transportation, and so on. The desired goal is to achieve equilibrium and stability, but this is not always attained. In heavy traffic where drivers follow each other very closely, an acceleration or deceleration of one vehicle may be a small disturbance that will be preserved or amplified in the system over time, suggesting that there can be sensitive dependence on initial conditions. This is a potential problem for traffic management, and can even result in accidents.

2.4.1 Car-following models

The starting point of a mathematical description of traffic flow problems is an equation describing the conservation of cars (cars are not created or destroyed). In one space dimension, this equation is a partial differential equation given by [26, 83, 242]

$$\rho_t + q_x = 0, \qquad (2.49)$$

where $\rho(x,t)$ is the density of cars and $q(x,t)$ represents the traffic flow (in physics: the "flux" of ρ across a boundary). However, for traffic problems where u is the average velocity of cars, $q = \rho u$. We make the simplified assumption that u depends only on the density of cars; that is, $u = u(\rho)$. This function can be determined experimentally (e.g., by counting the number of cars passing per hour). It may also be determined by using simple models. The motion of a line of vehicles on a crowded road link without overtaking (a rash assumption) is described by a car-following model [33]. This model is based on the assumption that a driver responds to the motion of the vehicle immediately in front. In the simplest model, the acceleration on the following car is assumed to be proportional to the difference between its speed and that of the car in front:

$$\frac{d^2x_n}{dt^2} = -\lambda\left(\frac{dx_n}{dt} - \frac{dx_{n-1}}{dt}\right). \tag{2.50}$$

If the car following is going faster than the preceding one, then the car following will slow down (and thus $\lambda > 0$). The larger the relative velocity, the more the car behind accelerates or decelerates. λ measures the sensitivity of the two-car interaction. However, Eq. (2.50) suggests that acceleration or deceleration occurs instantaneously. Instead, let us allow some time before the driver reacts to changes in the relative velocity. The process is modeled by specifying the acceleration at a slightly later time

$$\frac{d^2x_n(t+\tau)}{dt^2} = -\lambda\left(\frac{dx_n}{dt} - \frac{dx_{n-1}}{dt}\right), \tag{2.51}$$

where τ is the reaction time. Mathematically, this equation is a DDE. Integrating Eq. (2.51) once yields

$$\frac{dx_n(t+\tau)}{dt} = -\lambda(x_n - x_{n-1}) + d_n, \tag{2.52}$$

an equation relating the velocity of cars at a later time to the distance between cars. Imagine a steady-state situation in which all cars are equidistant apart, and hence moving at the same velocity. Thus

$$\frac{dx_n(t+\tau)}{dt} = \frac{dx_n(t)}{dt}, \tag{2.53}$$

and hence letting $d_n = d$

$$\frac{dx_n(t)}{dt} = -\lambda(x_n - x_{n-1}) + d. \tag{2.54}$$

Because

$$x_{n-1} - x_n = \frac{1}{\rho} \tag{2.55}$$

is a reasonable definition of traffic density, this model yields a velocity–density relationship

$$u = \frac{\lambda}{\rho} + d. \tag{2.56}$$

We choose the one arbitrary constant d such that at maximum density (bumper-to-bumper traffic) $u = 0$. In other words

$$0 = \frac{\lambda}{\rho_{\max}} + d. \tag{2.57}$$

In this way the following velocity–density relationship is derived,

$$u = \lambda \left(\frac{1}{\rho} - \frac{1}{\rho_{\max}} \right). \tag{2.58}$$

See Figure 2.7. How does this compare with experimental observations of velocity–density relationships? Equation (2.58) appears reasonable for large densities, that is, near $\rho = \rho_{\max}$. However, it predicts an infinite velocity at zero density. We can eliminate this problem by noting that this model is not appropriate for small densities for the following reasons. At small densities, the change of speed of a car is not due to the car in front. Instead it is more likely that the speed limit influences a car's velocity (and acceleration) at small densities. Thus we may hypothesize that Eq. (2.58) is valid only for large densities. For small densities, u is only limited by the speed limit $u = u_{\max}$ (see Figure 2.7).

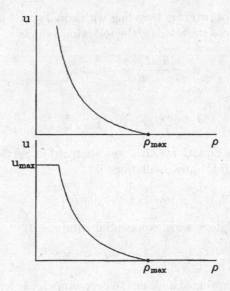

Figure 2.7: The speed u as a function of the density ρ. The top figure shows the hyperbolic function as predicted by the car-following model. In practice, the maximum permitted speed $u = u_{\max}$ is introduced (see bottom figure).

2.4.2 Local and asymptotic stability

When the lead vehicle of a line of cars changes its motion, the response of the following vehicle and the global response of all the cars in the line will not be the same. In this section we address this question by considering both the stability of two successive cars as well as the stability of a large numbers of cars.

Equation (2.51) can be solved by the method of Laplace transform [93] but the evaluation of the inverse Laplace transform may lead to a complex expression with little physical insight. In this section, we address the stability of the linear car following Eq. (2.51) with respect to disturbances. Two particular types of stabilities need to be examined, *local stability* and *asymptotic stability*. Local stability is concerned with the response of the following vehicle to a fluctuation in the motion of the vehicle directly in front of it; that is, it concentrates on the localized behavior between pairs of vehicles. Asymptotic stability is concerned with the manner in which a fluctuation in the motion of any vehicle, say the lead vehicle of a platoon, is propagated through a line of vehicles.

Local stability

From Eq. (2.51), we determine the equation for the velocity $v_n = dx_n/dt$ given by

$$\frac{dv_n(t+\tau)}{dt} = -\lambda(v_n - v_{n-1}).$$ (2.59)

Consider the case of two cars traveling with equal speed u. Assuming that the lead vehicle keeps its velocity, the following vehicle $v_n = u + y$ satisfies

$$\frac{dy(t+\tau)}{dt} = -\lambda y,$$ (2.60)

or equivalently,

$$\frac{dy}{ds} = -\lambda\tau y(s-1),$$ (2.61)

where $s = t/\tau + 1$. Equation (2.61) was analyzed in Section 2.1. We know that y does not exhibit any oscillations if

$$\lambda\tau \leq e^{-1} \simeq 0.37$$ (2.62)

and that y is oscillatory with exponential damping if

$$e^{-1} < \lambda\tau < \pi/2 \simeq 1.57.$$ (2.63)

In order for the following vehicle not to overcompensate for a fluctuation, it is necessary that condition (2.62) be verified. The criterion of local stability is often referred to this condition although the steady state is unstable only if $\lambda\tau > \pi/2$.

Asymptotic stability

Now assume that the lead driver's velocity varies periodically as

$$v_0 = 1 + \frac{1}{2}\left(\exp(i\omega t) + c.c\right). \qquad (2.64)$$

Also assume that the nth driver's velocity varies periodically

$$v_n = 1 + \frac{1}{2}\left(f_n \exp(i\omega t) + c.c\right), \qquad (2.65)$$

where f_n measures the amplification or decay that occurs. Starting with $f_0 = 1$, we determine f_1 as a function of f_0, then f_2 as a function of f_1, and so on. Iterating n times, the solution is given by

$$f_n = \frac{1}{\left[1 + \frac{i\omega}{\lambda}\exp(i\omega\tau)\right]} f_{n-1} = \frac{1}{\left[1 + \frac{i\omega}{\lambda}\exp(i\omega\tau)\right]^n} f_0. \qquad (2.66)$$

Thus, the amplitude $|f_n|^2$ is computed as

$$|f_n|^2 = \frac{1}{\left(1 + \frac{i\omega}{\lambda}\exp(i\omega\tau)\right)^n} \frac{1}{\left(1 - \frac{i\omega}{\lambda}\exp(-i\omega\tau)\right)^n} |f_0|^2$$

$$= \left[\frac{1}{\left(1 + \frac{\omega^2}{\lambda^2} - \frac{2\omega}{\lambda}\sin(\omega\tau)\right)}\right]^n. \qquad (2.67)$$

We next wish that $|f_n|^2 \to 0$ which means that

$$\frac{\omega^2}{\lambda^2} - \frac{2\omega}{\lambda}\sin(\omega\tau) > 0 \quad \text{or} \quad \frac{\omega}{\lambda} - 2\sin(\omega\tau) > 0 \quad \text{or} \quad \frac{\sin(\omega\tau)}{\omega} < \frac{1}{2\lambda}. \quad (2.68)$$

The inequality holds for all ω if [3]

$$\lambda\tau < \frac{1}{2}. \qquad (2.69)$$

We conclude that if the product of the sensitivity and the time lag is greater than 0.5, it is possible for following cars to drive more erratically than the leader. In this case, we say that the model predicts instability if $\lambda\tau > \frac{1}{2}$. Note that the criterion for local stability (namely that no local oscillations occur if $\lambda\tau < e^{-1}$) also insures asymptotic stability.

2.5 Bistability

Recent experiments on polarization switching in lasers subject to optical feedback [206] have motivated a simple analytical study of a first-order DDE. The laser (here a vertical-cavity surface-emitting laser) is subject to

---

[3] We investigate the function $\lambda = \omega/2\sin(\omega\tau)$ and note that the limit ω small leads to the stability limit.

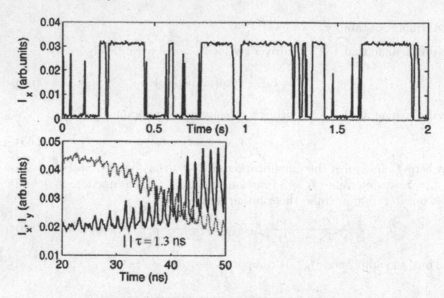

Figure 2.8: Top: Polarization mode-hopping. Bottom: Blow-up of the fast oscillatory jump transition between modes. The frequency $f \sim 450$ MHz is close to the inverse of the delay τ (from Sciamanna et al [206]).

optical feedback from a distant mirror (see Chapter 7). As a result, the light reinjected into the laser corresponds to the laser output at time $t - \tau$ where $\tau = 2L/c$. $L = 20.2$ cm being the distance laser-mirror and $c = 3 \times 10^8$ m/s the speed of light, we determine $\tau = 1.3$ ns. Compared to the time-scale of the laser (i.e., the photon lifetime $\tau_p \sim 1$ ps), this delay is large and we may reasonably expect some impact on the laser response. The experiments indicate that the laser exhibits polarization mode hopping due to noise (i.e., spontaneous emission noise) with fast oscillatory jump transitions with a period close to τ. See Figure 2.8. These transient oscillations are not specific to the laser but have been found by numerically investigating the following first-order DDE,

$$x' = x - x^3 + cx(t - \tau) + \sqrt{2D}\xi(t) \qquad (2.70)$$

where $\xi(t)$ is a Gaussian white noise of zero mean and unitary variance and D is the noise level. Equation (2.70) has been analyzed by Tsimring and Pikovsky [228] and Masoller [155] in the general context of a bistable system subject to noise. Their studies motivated further experimental work using a laser subject to a time-delayed optoelectronic feedback [101].

In this section, we illustrate the technique of linearization by examining the stability of the steady states of Eq. (2.70) with $D = 0$. It admits the following steady state-solutions

$$x = 0, \tag{2.71}$$

$$x = x_\pm \equiv \pm\sqrt{1+c}\ (c \geq -1). \tag{2.72}$$

From (2.72), we note that two non-zero steady states are branching from the zero solution at $c = -1$. Introducing the deviation $u = x - x_s$ where x_s is either (2.71) or (2.72), and assuming u small, we obtain a linear DDE. Looking then for exponential solutions leads to the following characteristic equation for the growth rate σ,

$$\sigma = 1 - 3x_s^2 + c\exp(-\sigma\tau). \tag{2.73}$$

As for Eq. (2.4), we examine this equation by first considering the case σ real and then the case σ complex ($\sigma = \sigma_r + i\sigma_i$). In the first case, we determine an implicit solution for $c = c(\sigma)$ which can be analyzed. In the second case, we formulate the parametric solution $\sigma_r = \sigma_r(\sigma_i\tau)$ and $c = c(\sigma_i\tau)$ which may or may not be expressed in terms of analytical functions.

If $c < -1$, $x_s = 0$ is the only steady state. It may change stability at a Hopf bifurcation provided τ is sufficiently large. Inserting $\sigma = i\sigma_i$ into (2.73) with $x_s = 0$, we obtain

$$1 - c\cos(\sigma_i\tau) = 0, \tag{2.74}$$

$$\sigma_i + c\sin(\sigma_i\tau) = 0. \tag{2.75}$$

Equivalently, we may formulate the parametric solution

$$c = \frac{1}{\cos(s)} \quad \text{and} \quad \tau = -\frac{s}{\tan(s)}, \tag{2.76}$$

where $s = \sigma_i\tau$. If $c > -1$, $x_s = 0$ always admits a real positive σ and is therefore unstable.

For the non–zero intensity steady-states $x_s = x_\pm$, a Hopf bifurcation is possible. Inserting $\sigma = i\sigma_i$ into (2.73), we now obtain

$$-2 + c(-3 + \cos(\sigma_i\tau)) = 0, \tag{2.77}$$

$$\sigma_i + c\sin(\sigma_i\tau) = 0 \tag{2.78}$$

from the real and imaginary parts. These conditions can be rewritten as

$$c = -\frac{2}{3 - \cos(s)} \quad \text{and} \quad \tau = \frac{s(3 - \cos(s))}{2\sin(s)}, \tag{2.79}$$

where $s = \sigma_i\tau$. Using (2.76) and (2.79), we may represent the Hopf bifurcation line in the (c, τ) stability diagram. See Figure 2.9. There are other Hopf bifurcation lines (not shown) that appear at higher values of c or τ.

The critical point $(c, \tau) = (-1, 1)$ is a degenerate Hopf bifurcation point because it corresponds to a double zero eigenvalue of the characteristic

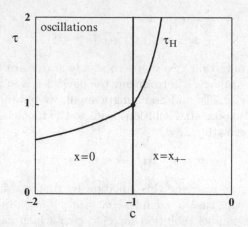

Figure 2.9: Stability diagram. The vertical line at $c = -1$ marks a steady bifurcation point from the zero to the non–zero steady state. The left curve connecting $(c, \tau) = (-1, 1)$ corresponds to a Hopf bifurcation point from the zero intensity steady state. The right curve starting at $(c, \tau) = (-1, 1)$ represents a Hopf bifurcation from the non–zero steady state. The critical point $(c, \tau) = (-1, 1)$ is a degenerate Hopf bifurcation point characterized by a double zero eigenvalue.

equation. The possible solutions near this point have been analyzed in detail by Redmond et al. [199] who derived a slow time second order differential equation for the small amplitude solutions. They showed that stable time-periodic solutions may coexist with stable steady states.

2.6 Metastability

The evolution to a long time steady-state solution can be oscillatory and very slow. The numerical solution of Eq. (2.70) is shown in Figure 2.10 starting from a square wave initial profile. Figure 2.10 shows sustained oscillations with a period close to the delay that disappear on the long time-scale. By tuning parameters, it is possible to observe the oscillations for longer periods. The phenomenon has been called *metastability*.

This phenomenon can be analyzed in detail using the following DDE,

$$\varepsilon x' = -x + f(x(t-1)), \qquad (2.80)$$

where $\varepsilon = \tau^{-1}$, $f(x) = -1$ if $x < 0$, $f(x) = 1$ if $x > 0$, and $f(0) = 0$ [3]. The numerical solution is shown in Figure 2.11. By constructing the solution in

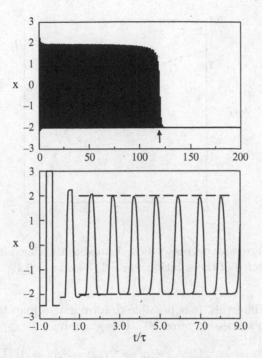

Figure 2.10: Top: slowly varying oscillations followed by a sudden jump to the steady-state $x = -2$. Bottom: short time solution showing the initial conditions: $x = -2.45$ $(-\tau < t < -2\tau/3$ and $-\tau/3 < t < 0)$ and $x = 3$ $(-2\tau/3 < t < -\tau/3)$. The values of the parameters are $c = 3$ and $\tau = 5$.

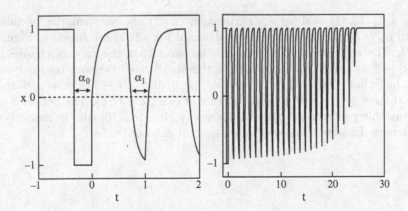

Figure 2.11: Right: The oscillations are slowly changing until they suddenly disappear and are replaced by the stable steady state $x = 1$. Left: Initial oscillations showing the first two intervals between successive zeros. $\varepsilon = 0.1$ and the initial function is $x = 1$ $(0 \leq t < 2/3)$, $x = -1$ $(2/3 \leq t < 0)$, and $x(0) = 0$.

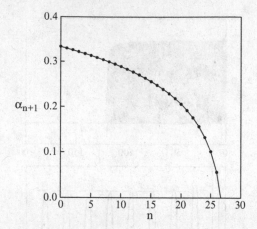

Figure 2.12: Progressive decrease of α_n defined as the time interval between two successive zeros of the solution. $\alpha_0 = 1/3$ and $\varepsilon = 0.1$.

successive time intervals, it is possible to formulate a map for the interval α_n between two successive zeros [81]. It is given by

$$\alpha_{n+1} = \alpha_n + \varepsilon \ln \left[\frac{2 - 2\exp(-\varepsilon^{-1}\alpha_n) + \exp(-\varepsilon^{-1})}{2 - \exp(-\varepsilon^{-1}(1 - \alpha_n))} \right] \tag{2.81}$$

which for small ε and $\alpha_0 < 0.5$ simplifies as

$$\alpha_{n+1} - \alpha_n \simeq \varepsilon \exp(-\varepsilon^{-1}\alpha_n) \tag{2.82}$$

The map (2.81) and its approximation (2.82) are mathematically valid until $\alpha_n = O(\varepsilon)$. The different iterations of Eq. (2.82) are shown in Figure 2.12. The expression (2.82) clearly indicates that the rate of change is $\varepsilon \exp(-\varepsilon^{-1})$ small. As noted in [81], this small rate of change results from the fact that $f(x(t-1)) = \pm 1$ for $x \gtrless 0$. If $f(x(t-1)) = a < 0$ and $b > 0$ for $x \gtrless 0$, and if $|a| \neq b$, the rate of change is $O(\varepsilon)$. The behavior of metastable patterns is further analyzed by Nizette [178] who formulated a Ginzburg–Landau equation from a general class of DDEs.

3
Biology

Many complex processes in biology and physiology are described by ordinary differential or functional differential equations. The latter are dominant when the functional components in equations allow us to consider after-effects of prehistory influence. Various classes of functional differential equations appear in immunology, epidemiology, and the theory of neural networks. Some areas are very well documented and benefit from regular reviews. This is, for example, the case of the glucose–insulin regulatory system [145], the cardiovascular control system [67], and blood pressure oscillations [200]. Here preference is given to problems that we easily understand such as population dynamics or postural control.

As described in the introduction, DDEs have quite long ago shown their efficiency in the study of the behavior of real populations. The majority of DDEs appearing as mathematical models of oscillatory biological phenomena are essentially nonlinear, as a rule, each of them having its own specific features. Because of this, we are often obliged to seek an individual approach to every concrete equation. The logistic DDE admits nearly harmonic oscillations that can be captured by a Hopf bifurcation analysis. On the other hand, the Mackey equation describing a strongly nonlinear self-inhibitory system requires a more subtle approach.

T. Erneux, *Applied Delay Differential Equations*, Surveys and Tutorials
in the Applied Mathematical Sciences 3, DOI 10.1007/978-0-387-74372-1_3,
© Springer Science+Business Media, LLC 2009

3.1   Population periodic cycles

The population densities of many species can fluctuate nearly periodically over time, with periods that cannot be explained simply by seasonal variation. These regular, large-amplitude oscillations have fascinated generations of ecologists [123]. In Chapter 1, we introduced the delayed logistic equation (1.21) as one of the simplest population models where a time lag appeared in the density-dependent term. The incorporation of the delay allows the description of sustained oscillations in a single species population, without any predatory interaction of other species.

Mathematically, we have determined a critical value of $\lambda = \lambda_0 = \pi/2$ above which the steady-state $y = 1$ is unstable and we have found by numerical simulations that sustained oscillations are possible if $\lambda > \lambda_0$. In Figure 3.1, we represent the extrema of the period of the oscillations as a function of λ. The oscillations emerge from the Hopf bifurcation point $\lambda = \lambda_0$. The broken line corresponds to the Hopf bifurcation approximation described in the next section. Near the bifurcation point, Hopf showed that

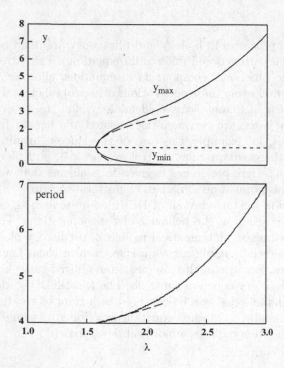

Figure 3.1: Bifurcation diagram of the steady and periodic solutions of the logistic DDE. Top: Extrema of the oscillations as a function of λ. Bottom: Period of the oscillations. The broken lines represent the Hopf bifurcation approximation.

the amplitude of the oscillations follows a square-root law (i.e., $y_{\max} - 1$ and $y_{\min} - 1$ are proportional to $(\lambda - \lambda_0)^{1/2}$) and the period changes linearly with $\lambda - \lambda_0$. A classical example of a comparison between experiments and theory comes from the experiments of the Australian ecologist, Nicholson [176], using laboratory cages of sheep blowflies, which can be a serious pest to sheep in Australia. Nicholson independently controlled the adult and larval food supplies and observed population oscillations of about 35 to 40 days. We use the delayed logistic equation with K set by the food level available and τ being the approximative time for a larva to mature into an adult. The actual value of K is not important because the period does not depend on K but only on $\lambda \equiv r\tau$. r is the rate of population increase and is unknown. Figure 3.2 compares the experiments with the solution of the logistic DDE for $\lambda = 2.1$. The period of the oscillations is about 4.54τ and if we consider the observed period of 40 days, we obtain a delay of 9 days. But the actual delay is 14 days. To overcome this discrepancy, Gurney et al. [82] proposed the following delay model (now referred as Nicholson's blowflies equation)

$$\frac{dN}{dt'} = rN(t' - \tau)\exp(-N(t' - \tau)/K) - mN. \tag{3.1}$$

Introducing the dimensionless variables (1.22) into Eq. (3.1), we obtain

$$\frac{dy}{dt} = ay(t - 1)\exp(-y(t - 1)) - by, \tag{3.2}$$

where

$$a \equiv r\tau \quad \text{and} \quad b \equiv m\tau. \tag{3.3}$$

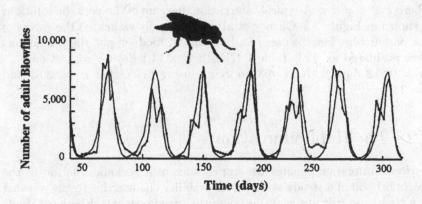

Figure 3.2: Blowfly numbers are fit with the solution of the logistic DDE for $\lambda = 2.1$ (after Nicholson [176] and May [153] redrawn from Case [45] p. 120).

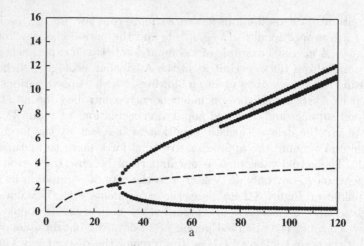

Figure 3.3: Bifurcation diagram of the extrema of the oscillations for $b = 3$. Note the appearance of a second maximum–minimum at $a > 70$. The broken line is the nonzero steady state. The Hopf bifurcation point is at $a = 29.69$.

The nonzero steady state is

$$y = \ln(a/b) \tag{3.4}$$

and its first Hopf bifurcation in terms of a and b is given in parametric form as

$$b = -\omega \cot(\omega), \tag{3.5}$$

$$a = b \exp(1 - \sec(\omega)), \tag{3.6}$$

where $\pi/2 < \omega < \pi$. A typical bifurcation diagram of the periodic solutions is shown in Figure 3.3. Gurney et al. [82] critically estimated the values of the parameters. For the case of adult limited food supply, the parameters were evaluated as [123] Table 1 (NLF): $\tau = 14.1$ days, $r = 5.93$ days^{-1}, $m = 0.212$ days^{-1}, $K = 463$ implying, using (3.3), that $a = 83.61$ and $b = 2.99$.

3.2   The Hopf bifurcation

A Hopf bifurcation denotes the appearance of a periodic solution in the neighborhood of a steady state whose stability changes due to the crossing of a conjugate pair of eigenvalues over the imaginary axis. Eberhard Hopf[1]

---

[1]Eberhard Hopf (1902–1983) is best known for his work in topology and ergodic theory. Born in Austria, he did most of his studies in Germany. In 1931, with the help of

formulated and proved his theorem about the appearance of periodic solutions for ordinary differential equations giving all credit to Poincaré [100]. The Hopf bifurcation theorem can be stated as follows.

Consider the differential equation

$$y' = F(y, \lambda), \tag{3.7}$$

where λ is a control parameter and suppose it admits a steady-state solution $y = y_0(\lambda)$. From the linearized problem,

$$u' = F_y(y_0(\lambda), \lambda)u, \tag{3.8}$$

we find a conjugate pair $\pm i\omega_0$ of simple pure imaginary eigenvalues at $\lambda = \lambda_0$. $\omega_0 > 0$ and no other eigenvalues exist on the imaginary axis. The second main hypothesis stated by Hopf is that the continuous extension to $\lambda - \lambda_0 \neq 0$ of the eigenvalue $i\omega_0$, say $\sigma(\lambda)$, should transversely cross the imaginary axis at $\lambda = \lambda_0$; that is,

$$\text{Re}(\sigma'(\lambda_0)) \neq 0. \tag{3.9}$$

The conclusion of the theorem then is that in a neighborhood of $(y, \lambda) = (y_0, \lambda_0)$, there always exists a one-parameter family $y(t, \varepsilon)$, $\lambda(\varepsilon)$ of periodic solutions (with $y(t, \varepsilon) \to y_0$, $\lambda \to \lambda_0$ as $\varepsilon \to 0$) having periods $T(\varepsilon)$ (with $T \to 2\pi/\omega_0$ as $\varepsilon \to 0$). When $F(y, \lambda)$ is analytic, ε can be chosen so that y, λ, and T are analytic, and

$$y = y_0 + \varepsilon y_1(t) + ..., \quad \lambda = \lambda_0 + \varepsilon^2 \lambda_2 + ..., \quad T = \frac{2\pi}{\omega_0}(1 + \varepsilon^2 T_2 + ...). \tag{3.10}$$

It turns out that $\lambda_2 \neq 0$, which is the generic situation; then for small ε, the periodic solution exists either *supercritically*, when $\lambda_2 \text{Re}(\sigma'(\lambda_0)) > 0$, or else *subcritically*, when $\lambda_2 \text{Re}(\sigma'(\lambda_0)) < 0$. If all the other eigenvalues of the linearized problem are in the left half-plane, Hopf showed that supercritical (subcritical) solutions are stable (unstable). See Figure 3.4.

In this section, we construct a periodic solution of a DDE by using the Hopf perturbation technique also known as Lindstedt-Poincaré method. It considers the sequential computation of power series solutions in an amplitude ε and uses solvability conditions as the most economic way to

Norman Wiener, Hopf became assistant professor at the Department of Mathematics at MIT. In 1936, at the end of the MIT contract, Hopf received an offer of full professorship at the University of Leipzig and moved back to Germany. In 1949, he became a U.S. citizen and joined Indiana University as a professor. Many people didn't forgive his moving to Germany. As a result most of his work on ergodic theory and topology was neglected or even attributed to others in the years following the end of World War II [250]. The Hopf bifurcation paper appeared in 1942. When asked many years later, he could not recall how he came to this problem but presumably it came through his interest in fluid mechanical instabilities [78].

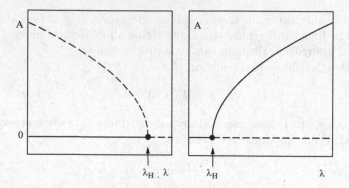

Figure 3.4: Left: In the case of a subcritical bifurcation, a branch of unstable periodic solutions overlaps a branch of stable steady states. Right: In the case of a supercritical Hopf bifurcation, a branch of stable periodic solutions overlaps a branch of unstable steady states.

compute the solution. Because supercritical solutions are stable and can be observed numerically, the determination of λ_2 is of particular interest. A convenient way to organize the calculation is shown for the delayed logistic equation (1.23). This equation is also considered in [111], p150, using the same technique, and in [51], p299, by the center manifold technique.

Equation (1.23) has the steady-state $y = 1$ and the linearized problem admits a pair of purely imaginary eigenvalues at $\lambda = \lambda_0 = \pi/2$. See (1.25). We now seek a 2π-*periodic solution* of Eq. (1.23) of the form

$$y - 1 = \varepsilon y_1(s) + \varepsilon^2 y_2(s) + ..., \tag{3.11}$$

where

$$s \equiv \omega t \tag{3.12}$$

is a scaled time variable. The parameter ε measures the amplitude of the periodic oscillations and can be defined by a normalization condition. A convenient definition is

$$\varepsilon \equiv \frac{1}{2\pi} \int_0^{2\pi} y(s, \varepsilon) \exp(-is) ds \tag{3.13}$$

implying

$$\frac{1}{2\pi} \int_0^{2\pi} y_1(s) \exp(-is) ds = 1 \quad \text{and} \quad \int_0^{2\pi} y_j(s) \exp(-is) ds = 0, \tag{3.14}$$

where $j = 2, 3, ...$ We next expand the bifurcation parameter λ and the frequency ω in power series of ε^2

$$\lambda = \lambda_0 + \varepsilon^2 \lambda_2 + ... \tag{3.15}$$

$$\omega = \omega_0 + \varepsilon^2 \omega_2 + ..., \tag{3.16}$$

where $\lambda_0 = \pi/2$ and $\omega_0 = 1$ are the values of the parameter λ and the frequency of the oscillations at the Hopf bifurcation point. The unknown coefficients λ_2, ω_2, ... will be obtained by applying solvability conditions at each order of the perturbation analysis. Inserting (3.11), (3.12), (3.15), and (3.16) into Eq. (1.23), we obtain from the first three orders appearing:

$$\omega_0 y_1' = -\lambda_0 y_1(s - \omega_0) \tag{3.17}$$

$$\omega_0 y_2' = -\lambda_0 y_2(s - \omega_0) - \lambda_0 y_1 y_1(s - \omega_0) \tag{3.18}$$

$$\omega_0 y_3' = -\lambda_0 y_3(s - \omega_0) - \lambda_0(y_1 y_2(s - \omega_0) + y_2 y_1(s - \omega_0))$$
$$- \omega_2 y_1' - \lambda_2 y_1(s - \omega_0) + \lambda_0 \omega_2 y_1'(s - \omega_0). \tag{3.19}$$

The solution of Eq. (3.17) satisfying the normalization condition (3.14) is given by

$$y_1 = \exp(is) + c.c., \tag{3.20}$$

where $c.c.$ means complex conjugate. The solution of Eq.(3.18) satisfying the normalization condition (3.14) is given by

$$y_2 = p_2 \exp(2is) + c.c., \tag{3.21}$$

where

$$p_2 = \frac{2-i}{5} \tag{3.22}$$

and there is no constant contribution. Solvability of Eq. (3.19) implies that there are no terms of the form $\exp(\pm is)$ in its right-hand side. This leads to the condition

$$-\lambda_0 \frac{(-1+3i)}{5} - i\omega_2 + i\lambda_2 + \lambda_0 \omega_2 = 0. \tag{3.23}$$

The real and imaginary parts give

$$\omega_2 = -\frac{1}{5} \quad \text{and} \quad \lambda_2 = \frac{3\pi - 2}{10}. \tag{3.24}$$

In summary, the Hopf leading asymptotic approximation of the solution is

$$y - 1 \simeq 2\sqrt{\frac{10(\lambda - \lambda_0)}{3\pi - 2}} \cos(\omega s), \tag{3.25}$$

where

$$\omega \simeq \omega_0 - \frac{2(\lambda - \lambda_0)}{3\pi - 2} \tag{3.26}$$

is the frequency. The extrema of the oscillations are given by

$$y - 1 = \pm 2\sqrt{\frac{10(\lambda - \lambda_0)}{3\pi - 2}} \tag{3.27}$$

and the period $P \equiv 2\pi/\omega$ depends on λ as

$$P = \frac{2\pi}{\omega_0}\left(1 + \frac{2(\lambda - \lambda_0)}{(3\pi - 2)\omega_0}\right). \tag{3.28}$$

Both approximations of the extrema and the period are shown in Figure 3.1 by broken lines.

3.3   Time-delayed negative feedback

3.3.1   Circulating red blood cells

An important scalar DDE exhibiting a time-delayed negative feedback frequently appears in the biological literature [63, 172]. This equation was originally proposed by Mackey [147] for an autoimmune disease that causes periodic crashes in circulating red blood cells (RBC). But it has also been used for other problems (see, e.g., [144]). The Mackey equation (best known as the Mackey-Glass equation [148]) exhibits limit-cycle oscillations but the delay is moderate. It is a strong negative feedback that is responsible for the oscillations.

When the RBC level in the blood is low, the cells produce a hormone, called erythropoietin, that stimulates the production of RBC precursor cells. After a few days, these precursors become mature RBCs and the production of erythropoietin is turned down. Hence, circulating RBC is controlled by a negative-feedback system with time-delay (maturation time). Mackey formulates a simple model given by [147]

$$\frac{dE}{dt'} = F(E(t' - \tau')) - \gamma E, \tag{3.29}$$

where E (cells/kg) is the circulating density of RBC, F (cell/(kg day)) is the cell influx from erythroid colony forming units, under erythropoietin control, τ' (days) is the time required to pass recognizable precursors, and γ (day^{-1}) is the loss rate of RBCs in the circulation. Experiments show that the feedback function saturates at low erythrocyte numbers and is a decreasing function of increasing red blood cell levels (negative feedback). Mackey chose to describe the feedback term by

$$F = F_0 \frac{\theta^p}{E^p(t' - \tau') + \theta^p}, \tag{3.30}$$

where F_0 (cell/kg day) is the maximal red blood cell production rate that the body can approach at very low circulating red blood cell numbers, p is a positive exponent, and θ (cell/kg) is a shape parameter. Parameters were estimated in the normal situation (not autoimmune hemolytic anemia) and are listed in the following table.

Parameters	Symbol	Values
RBC loss rate	γ	2.31×10^{-2} day^{-1}
Maximal RBC production rate	F_0	7.62×10^{10} cell kg^{-1}day^{-1}
Steepness	p	7.6
Shape	θ	2.47×10^{11} cell kg^{-1}
Feedback delay	τ'	5.7 days

Introducing the dimensionless variables

$$y = \frac{E}{\theta} \quad \text{and} \quad t = \frac{F_0 t'}{\theta} \tag{3.31}$$

into Eqs. (3.29) and (3.30), we find

$$\frac{dy}{dt} = \frac{1}{1 + y(t-\tau)^p} - by, \tag{3.32}$$

where

$$\tau = \frac{F_0 \tau'}{\theta} \quad \text{and} \quad b = \frac{\gamma\theta}{F_0}. \tag{3.33}$$

The next table provides their values

Parameters	Symbol	Values
RBC loss rate	b	7.45×10^{-2}
Steepness	p	7.6
Delay	τ	1.73

Periodic hemolytic anemia can be induced in rabbits by administration of RBC autoantibodies. The immune system destroys RBCs, thereby increasing b. We thus treat b as our bifurcation parameter.

We note that the value of τ is an $O(1)$ quantity but that the steepness parameter $p = 7.6$ can be considered as relatively large. A limit-cycle solution of Eq. (3.32) for $p = 20$ is shown in Figure 3.5. Increasing p leads to triangular oscillations connecting two distinct exponential decaying functions of t. As noted by Mackey [147], the nonlinear function $[1 + y(t-\tau)^p]^{-1}$ in Eq. (3.32) approaches the Heavyside function as $p \to \infty$ allowing a drastic simplification of the nonlinear equation. The large p limit of Eq. (3.32) is linear as

$$\frac{dy}{dt} = -by + \begin{array}{ll} 0 & \text{if } y(t-\tau) > 1 \\ 1 & \text{if } y(t-\tau) < 1 \end{array}. \tag{3.34}$$

In Figure 3.5, $y(t-\tau) > 1$ during the time interval $0 < t < t_m$ and $y(t-\tau) < 1$ during the time interval $t_m < t < P$. We construct the solution

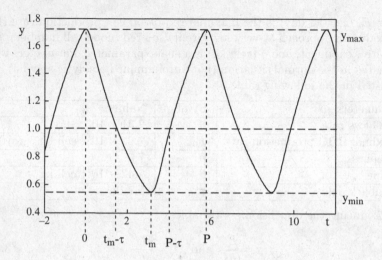

Figure 3.5: Limit-cycle solution. The values of the parameters are $b = 0.4$, $p = 20$, and $\tau = 1.8$.

for each time interval and connect them at the critical times $t = t_m$ and $t = P$. There are four steps in this analysis:

1. During the interval $0 < t < t_m$, the nonlinear function is zero and Eq. (3.32) is linear. With the initial condition $y(0) = y_{\max}$, it admits the solution

$$y = y_{\max} \exp(-bt). \tag{3.35}$$

2. At the critical time $t = t_m$, $y(t_m - \tau) = 1$ and $y(t_m) = y_{\min}$. Using (3.35), these two conditions imply

$$y(t_m - \tau) = 1 = y_{\max} \exp(-b(t_m - \tau)), \tag{3.36}$$

$$y_{\min} = y_{\max} \exp(-bt_m). \tag{3.37}$$

Eliminating $y_{\max} \exp(-bt_m)$ between (3.36) and (3.37) leads to a simple expression for $y_{\min}$ given by

$$y_{\min} = \exp(-b\tau). \tag{3.38}$$

3. During the interval $t_m < t < P$, the nonlinear function equals one and Eq. (3.34) is again linear. With the condition $y(t_m) = y_{\min}$, it admits the solution

$$y = (y_{\min} - b^{-1}) \exp(-b(t - t_m)) + b^{-1}. \tag{3.39}$$

4. At time $t = P$, $y(P - \tau) = 1$ and $y(P) = y_{\max}$. These conditions imply

$$y(P - \tau) = 1 = (y_{\min} - b^{-1}) \exp(-b(P - \tau - t_m)) + b^{-1}, \tag{3.40}$$

$$y_{\max} = (y_{\min} - b^{-1}) \exp(-b(P - t_m)) + b^{-1}. \tag{3.41}$$

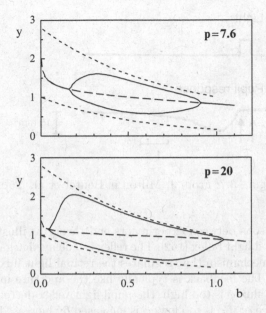

Figure 3.6: Effect of p. Bifurcation diagram of the extrema of y for the Mackey equation. $\tau = 1.73$. The middle line is the steady-state. The upper and lower broken lines are the approximations of the extrema for p large.

Eliminating $(y_{\min} - b^{-1}) \exp(-b(P - t_m))$, we obtain $y_{\max}$ as

$$y_{\max} = (1 - b^{-1}) \exp(-b\tau) + b^{-1}. \tag{3.42}$$

The expressions (3.38) and (3.42) for the extrema of the oscillations are compared to the numerical solutions in Figure 3.6. The top figure corresponds to the value of p estimated by Mackey. The two Hopf bifurcation points located at $b = 0.16$ and $b = 0.92$ limit a domain of unstable steady states. The estimated value of $b = 7.45 \times 10^{-2}$ for the normal RBC circulation is clearly below the first Hopf point and corresponds to a stable steady state. The bottom figure corresponds to a large value of p and we note a better agreement between numerical and analytical values except near the two Hopf bifurcation points ($b = 0.05$ and $b = 1.04$). As $p \to \infty$, the Hopf bifurcation becomes more and more vertical near $b = 0$ and $b = 1$, respectively. The analysis of the bifurcation diagram near $b = 0$ and $b = 1$ for p large is possible but is not a routine application of singular perturbation techniques.

3.3.2   Pupil light reflex

The idea of using piecewise constant feedback goes beyond the approximation of a strong negative feedback. Piecewise constant feedback allows

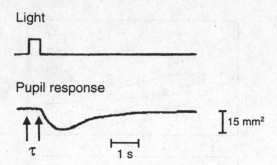

Figure 3.7: From J. Milton in Beuter et al. [25].

precise comparisons between experiments and theory as illustrated by the study of the pupil light reflex [162]. The reflex is a delayed negative feedback neural control mechanism that regulates the retinal light flux by changing the pupil area. The feedback is typically like the aperture in a camera: if the retinal light flux ϕ is too high, the pupil light reflex decreases the pupil area A and hence ϕ; if ϕ is too low, ϕ is increased by increased A. However, the feedback is time-delayed: pupil size does not change immediately in response to a change in illumination, but starts to react after a delay τ of about 300 msec. See Figure 3.7. If the gain and/or τ in the feedback loop becomes sufficiently large, sustained oscillations in the pupil area with a period T, where $2\tau < T < 4\tau$, were predicted theoretically. Direct experimental verification has been facilitated by the development of techniques to "clamp" the pupil light reflex. Clamping refers to techniques in which the feedback loop is first "opened" and then "reclosed." The main interest of this technique, referred to as piecewise constant negative feedback, is that it allows an analytical insight of the delay-induced oscillations. A recent review of the problem and its mathematical modeling is given by Milton in [25]. Under normal physiological conditions, we do not expect to see spontaneous regular oscillations in A. However, the gain in the pupil light reflex can be easily increased by focusing a narrow light beam at the pupillary margin. Despite the simplicity of this technique, systematic observations are difficult because many patients are distracted by the positioning of the light beam at the edge of their pupil. To overcome this difficulty, a known feedback is inserted into the reflex arc in a noninvasive way (clamping). The pupil area oscillations are then controlled experimentally using piecewise constant types of feedback: the light is either on or off depending on the value of the pupil area relative to certain area thresholds. A minimal model equation describing the pupil area change has been studied in detail by Milton and Longtin [163]. It has the form

$$\alpha^{-1}\frac{dA}{dt} + A = \begin{array}{ll} A_{on} & \text{if } A(t-\tau) > A_{ref} \\ A_{off} & \text{otherwise,} \end{array} \qquad (3.43)$$

where the right-hand side corresponds to piecewise constant feedback. As for Eq. (3.36) we may construct an analytical solution of the long-time periodic solution by connecting two distinct solutions. Equation (3.43) is slightly different from Eq. (3.34) because $A_{off} \neq 0$, $A_{on} \neq 1$, and $A_{ref} \neq 1$. Moreover, the rate constants for pupil movements significantly differ for constriction ($\alpha = \alpha_c$) and dilation ($\alpha = \alpha_d$). The delay times for constriction and dilatation are taken equal ($\tau_c = \tau_d = \tau$). The extrema of the oscillations and the period are given by

$$A_{\min} = A_{on} + (A_{ref} - A_{on})\exp(-\alpha_c\tau), \tag{3.44}$$

$$A_{\max} = A_{off} + (A_{\min} - A_{off})\exp(-\alpha_d(P - t_m)), \tag{3.45}$$

and

$$P = 2\tau + \frac{1}{\alpha_c}\ln(\frac{A_{\max} - A_{on}}{A_{ref} - A_{on}}) + \frac{1}{\alpha_d}\ln(\frac{A_{\min} - A_{off}}{A_{ref} - A_{off}}). \tag{3.46}$$

The values of the parameters are listed in the following table[2] and the comparison with the experiments is shown in Figure 3.8.

Parameters	Symbol	Value
Total delay	τ	380 ms
Rate of pupil contriction	α_c	4.46 s^{-1}
Rate of pupil dilation	α_d	0.42 s^{-1}
Initial pupil area	A_{on}	11.8 mm^2
Minimum pupil area after a 2 s light pulse	A_{off}	34.0 mm^2

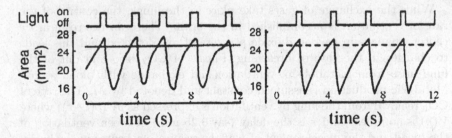

Figure 3.8: Experiment (left) and theory (right). $A_{ref} = 25$ mm^2 (redrawn from J. Milton in Beuter et al. [25]).

---

[2]The parameters were estimated from pupil cycling experiments. The total delay time is the neural time delay (280 ms) + machine time delay (100 ms). From Subject A, Table 2 in Milton and Longtin [163].

3.3.3 Periodic breathing

Mathematical models of the human respiratory control system have been developed since 1940 [65, 122, 18, 127]. The phenomena collectively referred to as periodic breathing or PB (including Cheyne–Stokes respiration and apneustic breathing) have important medical implications. In this section, we consider a purely quantitative approach of PB that allows treatment using widely understood clinical concepts.

A cyclic breathing pattern characterized by a smooth rise and fall in ventilation with cycle lengths ranging from ~25 to 100 s ($10^{-2} - 4 \times 10^{-2}$ Hz) is frequently observed in chronic heart failure (CHF) patients and is commonly referred to as periodic breathing (PB), or usually when separated by apnea, Cheyne–Stokes respiration [191]. Often, the same patient may exhibit a continuum of different patterns of breathing, ranging from normal breathing to mild PB up to cyclic periods of apnea. These patterns are also influenced by wakefulness or sleep, posture, and physical and mental activity. The physiological mechanisms responsible for PB in CHF patients are still a matter of debate. Two major hypotheses, however, have received the most attention in the last two decades. The "central" hypothesis explains PB as the result of a central vasomotor rhythm that modulates ventilation either indirectly through modulation of blood flow or directly through central irradiation to respiratory centers. The "instability" hypothesis, on the contrary, explains PB as a self-sustained oscillation due to loss of stability in the close loop chemical control of ventilation. The instability hypothesis has gained wider acceptance mainly because of its sound theoretical basis using mathematical models of the respiratory control system. These studies have shown that increased circulatory delay and loop gain brought about by the decreased cardiac output of CHF patients may lead to instability in their feedback control of ventilation.

While the exchange of gases take place in the lungs, the control of the rate of ventilation is accomplished in the brain. There, in the respiratory center, changes in carbon dioxide pressure of are detected and this leads to changes in the rate of breathing. Fluxes of CO_2 into and out of the lung arise from metabolism, ventilation, and exchange with blood stores. Metabolic production is assumed constant and denoted by M. The rate of CO_2 removal from the lung by ventilation is expressed as $pV(p(t-\tau))$ where $V(p)$ is nonlinear and τ is the delay (say 0.25 min) between ventilation of the blood and the measurement of p at the respiratory center in the brain. The transport of blood from the lungs back to the heart and then to the brain takes time. Oscillations in arterial CO_2 necessitate a net transfer of CO_2 from the lung into extrapulmonary stores and are described by $W(p)$. The rate of increase of lung CO_2 stores is $V_L dp/dt$ where V_L is the alveolar volume and it satisfies

$$V_L \frac{dp}{dt} = M - pV(p(t - \tau)) - W(p). \tag{3.47}$$

We assume that there exists a steady-state solution $p = p_0$. The linearized equation for the small deviation $u = p - p_0$ then is

$$V_L \frac{du}{dt} = u(V(p_0) + W'(p_0))u - p_0 V'(p_0)u(t - \tau). \qquad (3.48)$$

The coefficients appearing in (3.48) are all measurable quantities [68]. $V(p_0)$ is the mean value of alveolar ventilation. $W'(p_0) = \beta Q$ where β is the solubility of CO_2 in blood and Q is the cardiac output. $V'(p_0) = S$ is called the chemoreflex gain. Substituting $u = u_0 \exp(\sigma t)$ leads to the characteristic equation

$$\sigma \tau + x + y \exp(-\sigma \tau) = 0 \qquad (3.49)$$

where the coefficients x and y are defined by

$$x \equiv \frac{(V(p_0) + W'(p_0))\tau}{V_L} \quad \text{and} \quad y \equiv \frac{p_0 V'(p_0)\tau}{V_L}. \qquad (3.50)$$

PB is considered as a oscillatory instability resulting from a Hopf bifurcation. Substituting $\sigma = i\omega$ into (3.49), we obtain the following conditions from the real and imaginary parts:

$$x + y \cos(\omega \tau) = 0, \qquad (3.51)$$
$$\omega \tau - y \sin(\omega \tau) = 0. \qquad (3.52)$$

A parametric solution for $y = (x)$ is possible and is given by

$$y = \frac{\omega \tau}{\sin(\omega \tau)} \text{ and } x = -\omega \tau \cot(\omega \tau). \qquad (3.53)$$

Clinical measurements of x and y are shown in Figure 3.9. The almost straight line that separates patients with PB and patients without PB is given by (3.53) ($0.5\pi < \omega \tau < 0.9\pi$). For the 10 patients with PB, the observed cycle time averaged 1.2 ± 0.2 min and using $P/\tau = 2\pi/\omega \tau$, the delay is about 0.3 to 0.5 min.

3.3.4 Genetic oscillations

Gene expression is regulated by transcription factors which themselves are gene products. The resulting system of genetic interactions are interconnected dynamical systems, which generally generate oscillatory instabilities. Genetic oscillations are observed in a growing number of systems.

The best-studied examples of genetic oscillators are circadian clocks [53] and cell cycle oscillators [230]. More recently, oscillations involved in the segmentation of vertebrates [193] and oscillations of the tumor suppressor p53 [17] and of the NFκ–IκBα signaling module [97] have received

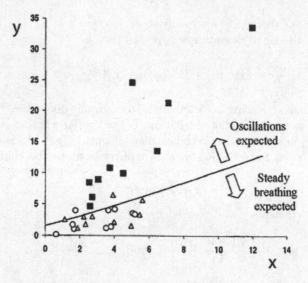

Figure 3.9: Observed values of x and y in patients with PB (squares) patients without PB (triangles), and normal controls (circles). The line is the Hopf boundary line (data taken from Francis et al. [68]).

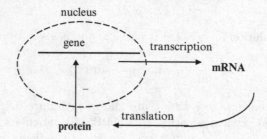

Figure 3.10: Delayed Hes1 negative feedback loop. The first process is the transcription and the export of the hes1 mRNA from the nucleus to the cytoplasm. The second process is the synthesis of Hes1 protein by translation of the hes1 mRNA. Both the transcription and the translation contribute to delay the feedback. Finally, the inhibition of the transcript initiation from the hes1 gene is controlled by the Hes1 protein.

attention. In order to understand the role of oscillatory gene expression, the study of simple models has proven to be extremely valuable [77]. The simplest case of a feedback oscillator is represented by a gene, its product and the corresponding mRNA (see Figure 3.10). If the gene product inhibits transcription of mRNA, the gene expression can oscillate if the time between the beginning of transcription and the end of translation can be represented by a time delay in the self-inhibitory system [169]. This mechanism has been proposed to provide the basis of segmentation oscillations [141]. Furthermore the p53-Mdm2 and the NFκB–IκBα systems can successfully be described using the same ideas. Oscillators based on autoinhibition with time delay possess the property that the oscillations period is mainly determined by the time delay and depends only weakly on the average protein expression rates [169].

Here, we review the analysis by Monk [169] for the oscillatory expression of Hes1 protein. The model is further developed in [23]. The analysis is based on the experiments by Hirata et al. [95] who observe that the transcription of messenger RNAs (mRNAs) for notch signaling molecules exhibits a two-hour cyclic behavior. Monk [169] and Jensen et al. [114] independently proposed to take into account the relatively large time delay for transport between the cell nucleus and the cytoplasm. Here the delay τ represents the lengthy processes of translation, transcription, and so on. See Figure 3.10. Denoting the concentrations of Hes1 mRNA by M and Hes1 protein by P, the minimal system of equations is given by

$$\frac{dM}{dt} = \alpha_m G(P(t - \tau)) - \mu_m M, \qquad (3.54)$$

$$\frac{dP}{dt} = \alpha_p M - \mu_p P, \qquad (3.55)$$

where μ_m and μ_p are the rates of degradation of mRNA and protein, respectively. α_m is the rate of transcript initiation in the absence of Hes1 protein, α_p is the rate at which Hes1 protein is produced from hes1 mRNA, and $G(P(t - \tau))$ is a monotonic decreasing function representing the delayed repression of hes1 mRNA production by Hes1 protein. G takes the general form

$$G(P) = \frac{1}{1 + (P/P_0)^n}, \qquad (3.56)$$

where P_0 is called the repression threshold and n is a Hill coefficient that determines the steepness of G. The delay τ represents the sum of the transcriptional and translational time delays. We may reduce the number of independent parameters by introducing the following dimensionless variables

$$m \equiv \mu_m \frac{M}{\alpha_m}, \quad p \equiv \frac{P}{P_0}, \quad \text{and} \quad s \equiv \mu_m t. \qquad (3.57)$$

Inserting (3.57) into (3.54) and (3.55), we obtain

$$\frac{dm}{ds} = \frac{1}{1 + p(s - \theta)^n} - m, \qquad (3.58)$$

$$a\frac{dp}{ds} = bm - p, \qquad (3.59)$$

where

$$\theta \equiv \mu_m \tau, \quad a \equiv \mu_m/\mu_p, \quad \text{and} \quad b \equiv \frac{\alpha_p \alpha_m}{\mu_p \mu_m} \frac{1}{P_0}. \qquad (3.60)$$

The values of the parameters considered by Monk [169] and the values of the dimensionless parameters θ, a, and b are listed in the following table.

Parameter	Value
$\mu_m = \mu_p$	0.03 min^{-1}
τ	18.5 min
$P_0/(\alpha_m \alpha_p)$	100 min^2
n	5
θ	0.56
a	1
b	11.11

The steady-state solution $m(b)$ and $p(b)$ is given in implicit form as

$$b = p(1 + p^n) \quad \text{and} \quad m = (1 + p^n)^{-1}. \tag{3.61}$$

From the linearized equations, we then determine the characteristic equation for the growth rate σ. It has the form

$$\sigma^2 + \sigma(1 + a^{-1}) + a^{-1}\left(1 + \frac{np^n}{1 + p^n}\exp(-\sigma\theta)\right) = 0, \tag{3.62}$$

where we eliminated b using the first steady-state expression in (3.61). The bifurcation parameter now is p instead of b. Inserting $\sigma = i\omega$ into (3.62), we obtain the two Hopf conditions from the real and imaginary parts. They are given by

$$-\omega^2 + a^{-1}\left(1 + \frac{np^n}{1 + p^n}\cos(\omega\theta)\right) = 0, \tag{3.63}$$

$$\omega(1 + a^{-1}) - a^{-1}\frac{np^n}{1 + p^n}\sin(\omega\theta) = 0. \tag{3.64}$$

Keeping a fixed, we seek an analytical solution for the Hopf boundary in the p versus θ plane. To this end, we introduce the rescaled frequency $x \equiv \omega\theta$ and eliminate $np^n/(1 + p^n)$ between the two equations. This leads to a quadratic equation for θ of the form

$$\theta^2 \tan(x) + \theta x(1 + a) - ax^2 \tan(x) = 0. \tag{3.65}$$

Having θ as a function of x, we use Eq. (3.64) to determine p. Solving for p^n, we obtain

$$p^n = \frac{x(1 + a)}{(n\theta\sin(x) - x(1 + a))}. \tag{3.66}$$

The Hopf bifurcation boundary is shown in Figure (3.11) for $a = 1$. There is a minimal value of θ below which a Hopf bifurcation is not possible. A vertical asymptote in the b versus θ stability diagram corresponds to the

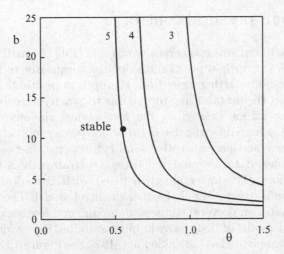

Figure 3.11: Hopf stability boundaries for $a = 1$ and $n = 5$, 4, and 3. The dot is located at $(\theta, b) = (0.555, 11.11)$.

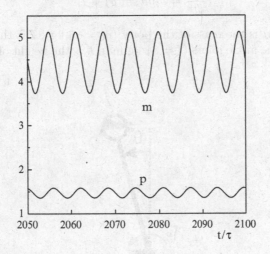

Figure 3.12: Periodic solution for $n = 5$, $\theta = 0.56$, $a = 1$, and $b = 11.11$. These values correspond to an unstable steady state located close to the Hopf stability boundary.

limit $b \to \infty$ or $q^n \to \infty$. From Eqs. (3.63) and (3.64) with $a = 1$, we determine

$$\omega_\infty = \sqrt{n - 1} \quad \text{and} \quad \theta_\infty = \frac{1}{\sqrt{n - 1}} \arcsin(\frac{2\sqrt{n - 1}}{n}). \qquad (3.67)$$

Figure 3.12 shows the periodic solution.

3.4   Human postural control

Upright stance in humans is inherently unstable [186]. A small sway deviation from a perfect upright position results in a torque due to gravity that
accelerates the body farther away from the upright position. To maintain
upright balance, the destabilizing torque due to gravity must be countered
by a corrective torque exerted by the feet against the support surface.
This control is achieved by the detection of body-sway motion through vi-
sual, vestibular,[3] and proprioceptive sensory[4] systems. In the laboratory,
stimulus–response data are analyzed using spectral analysis to compute
transfer functions. The latter are then fitted with numerically obtained
transfer functions using a simple feedback control model. The model then
provides estimates on postural stiffness, damping, and feedback time delay.

The simplest model of body sway in quiet standing is provided by an inverted pendulum pivoted at the ankle joint [187]. See Figure 3.13. Balancing
the torques provides the biomechanical model for the body sway θ given by

$$I\frac{d^2\theta}{dt^2} = mgh\sin(\theta) + T_c. \qquad (3.68)$$

In Eq. (3.68), m is the mass of the body ($m = 85$ kg), I is the moment of
inertia about the ankle joints ($I = 81$ kg m^2), h is the height of center mass

Figure 3.13: Feedback model of the eyes closed postural control system.
The body is modeled as an inverted pendulum. Body position in space (θ)
is sensed by the graviceptive system. A neural controller models the corrective torque T_c generated in response to the internal orientation estimate
y (Fig. 2 modified with permission from Peterka and Loughtin [187]).

---

[3]The vestibular system (or balance system) is a sensory system that provides information on our movement and orientation in space. Together with the cochlea, the
auditory organ, it is situated in the vestibulum in the inner ear.

[4]Proprioception is the sense of the relative position of neighboring parts of the body.

above the ankle joint axis ($h = 0.9$ m), and T_c is the corrective torque about the ankle joint. The internal body-orientation estimate y is provided by the graviceptive sensory system and has the form $y = -W\theta$ where the weight $W(t)$ can vary in time but is typically close to 1 for eye-closed stance. The neural controller that takes into account both position and velocity information produces the corrective torque T_c as a function of y at time $t-t_d$. It can be described as $T_c = K_P y(t-t_d) + K_D dy(t-t_d)/dt + K_I \int y(t-t_d)dt$ where $K_P = 970$ N m rad^{-1}, $K_D = 344$ N m s rad^{-1}, $K_I = 86$ N ms^{-1}rad^{-1}, and $t_d = 0.175$ s were estimated from experimental data [187, 163]. Introducing the dimensionless time $s = \sqrt{mgh/It}$, Eq. (3.68) simplifies as

$$\frac{d^2\theta}{ds^2} = \sin(\theta) + M, \qquad (3.69)$$

where

$$M = -a\theta(s-\tau) - b\frac{d\theta(s-\tau)}{ds} - c\int \theta(s-\tau)ds \qquad (3.70)$$

and

$$a = \frac{K_P}{mgh}W, \ b = \frac{K_D}{\sqrt{mghJ}}W, \ c = \frac{K_i}{\sqrt{(mgh)^3/J}}, \ \text{and} \ \tau = \sqrt{mgh/I}t_d. \qquad (3.71)$$

Using the values of the parameters listed above, we find $a = 1.29$ W rad^{-1}, $b = 1.32$ W rad^{-1}, $c = 0.04$ W rad^{-1}, and $\tau = 0.53$. Because $c \ll a, b$, we may neglect the integral contribution to the corrective torque.

We next analyze the stability of the equilibrium position $\theta = 0$ by linearizing Eqs. (3.69) and (3.70) with $c = 0$. Our objective is to determine the region of stability in the a versus τ diagram. The characteristic equation for the growth rate σ is

$$\sigma^2 = 1 - a\exp(-\sigma\tau) - b\sigma\exp(-\sigma\tau) \qquad (3.72)$$

The case of purely real eigenvalues can be treated analytically by studying the inverse function $a = a(\sigma)$

$$a = (1 - \sigma^2)\exp(\sigma\tau) - b\sigma. \qquad (3.73)$$

From this analysis, we find that $\sigma < 0$ provided that $a > 1$. We next consider the case of a Hopf bifurcation. The conditions for a Hopf bifurcation are obtained by inserting $\sigma = i\omega$ into Eq. (3.72). From the real and imaginary parts, we find

$$\omega^2 + 1 - a\cos(\omega\tau) - b\omega\sin(\omega\tau) = 0, \qquad (3.74)$$

$$a\sin(\omega\tau) - b\omega\cos(\omega\tau) = 0. \qquad (3.75)$$

From the last equation, we find a as

$$a = b\omega\cot(\omega\tau) \qquad (3.76)$$

and eliminating a in (3.74) gives

$$\omega^2 + 1 - b\omega / \sin(\omega\tau) = 0. \qquad (3.77)$$

Equations (3.76) and (3.77) provide a solution for $a = a(\tau)$ in parametric form

$$\tau = \frac{1}{\omega} \arcsin(\frac{b\omega}{1+\omega^2}) > 0, \qquad (3.78)$$

$$a = \sqrt{(1+\omega^2)^2 - b^2\omega^2}. \qquad (3.79)$$

See Figure 3.14.

For $\tau = 0.53$, the equilibrium position $\theta = 0$ is stable if $1 < a < 2.2$. If we consider the Hopf bifurcation frequency at $\tau = 0.53$ ($\omega \simeq 1.3$) as an estimate of the oscillatory frequency, we find $f = 0.63$ Hz ($f = \omega\sqrt{mgh/I}/2\pi$) which is in the 1 Hz range observed experimentally.

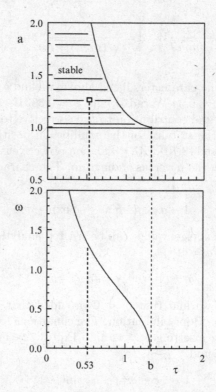

Figure 3.14: Stability region delineated by the Hopf bifurcation $\sigma = \pm i\omega$ and the line where $\sigma = 0$. The critical point $(\tau, a) = (b, 1)$ corresponds to a double zero eigenvalue ($\omega = 0$). The square marks the experimental point $(\tau, a) = (0.53, 1.29)$.

3.5   The inverted pendulum

Physically related to the postural control problem, the control of an inverted pendulum is a classic problem that has applications to both biological and mechanical balancing tasks. Stick-balancing experiments are a noninvasive method to learn how the nervous system reacts when a person acquires new skills and how it becomes accustomed to a specific task.

Cabrera and Milton [39] investigated experimentally the vertical displacement of a stick balanced at the fingertip. The observations are reproduced by a simple model for an inverted pendulum based on Newton's laws. A time-delayed feedback with parametric noise describes the effect of the neural control of balance (see Stépán and Kollár [214] who used a Lagrangian approach). Specifically, an inverted pendulum of mass m, length l, and moment of inertia $I = ml^2/3$ is subject to three forces namely, its weight, friction, and a restoring force applied by the hand, that depends on the angular deviation at time $t - \tau$. Balancing torques, the equation of motion (without noise) is of the form

$$I\theta'' + \gamma l^2\theta' - mgl\sin(\theta) + r_0 l\theta(t - \tau) = 0, \qquad (3.80)$$

where prime means differentiation to time t. The minus sign multiplying $\sin(\theta)$ is because we have taken $\theta = 0$ to be the upright position. Introducing $s \equiv t/\tau$, Eq. (3.80) can be rewritten in dimensionless form as

$$\theta'' + \Gamma\theta' - q\sin(\theta) + R_0\theta(s - 1) = 0, \qquad (3.81)$$

where prime now means differentiation with respect to time s. The dimensionless coefficients are defined by

$$\Gamma \equiv \frac{3\gamma\tau}{m}, \quad q \equiv \frac{3g\tau^2}{l}, \quad \text{and} \quad R_0 \equiv r_0\frac{3\tau^2}{ml}. \qquad (3.82)$$

In addition to the zero solution, Eq. (3.81) admits a nonzero steady state given by (in implicit form)

$$R_0 = \frac{q\sin(\theta)}{\theta}. \qquad (3.83)$$

We next investigate the stability of the equilibrium position $\theta = 0$. From the linearized equation, we determine the following characteristic equation for the growth rate σ

$$\sigma^2 + \Gamma\sigma - q + R_0\exp(-\sigma) = 0. \qquad (3.84)$$

There exist two distinct stability boundaries corresponding to either $\sigma = 0$ or $\sigma = i\omega$. In the first case, we find from (3.84) $R_0 = q$ ($r_0 = mg$).

It corresponds to the bifurcation point to the nonzero steady-state solution (3.83). The second case implies the two conditions

$$-\omega^2 - q + R_0 \cos(\omega) = 0, \tag{3.85}$$

$$\Gamma\omega - R_0 \sin(\omega) = 0. \tag{3.86}$$

We wish to investigate these equations in terms of the feedback rate r_0 and delay τ. Using (3.82), Eqs. (3.85) and (3.86) become

$$-\omega^2 - \frac{3g\tau^2}{l} + \frac{3\tau^2 r_0}{ml} \cos(\omega) = 0, \tag{3.87}$$

$$\gamma\omega - \frac{r_0\tau}{l} \sin(\omega) = 0. \tag{3.88}$$

Using (3.88), we eliminate r_0 in Eq. (3.87) and obtain a quadratic equation for τ

$$\frac{3g}{l}\tau^2 - \gamma\omega\frac{3}{m}\cot(\omega)\tau + \omega^2 = 0. \tag{3.89}$$

Solving for $\tau = \tau(\omega)$, we then compute $r_0 = r_0(\omega)$ from (3.88) as

$$r_0 = \frac{\gamma l\omega}{\sin(\omega)} \frac{1}{\tau}. \tag{3.90}$$

The stability boundaries are shown in Figure 3.15. We have not analyzed the stability of the nonzero steady states. Considering $\tau = 0.07$ as in [39], we have verified numerically that the upper boundary is a supercritical

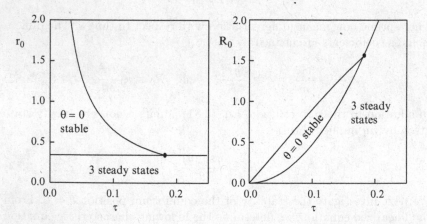

Figure 3.15: Stability boundaries. Left: Stability domain of $\theta = 0$ in terms of r_0 and τ. Right: Same stability domain in terms of $R_0 \equiv 3r_0\tau^2/(ml)$ and τ as shown in [39]. The values of the parameters are: $l = 0.62$ m, $m = 0.035$ kg, $\gamma = 0.1$ kg s^{-1}, and $g = 9.81$ ms^{-2}. The thin line corresponds to a bifurcation point from the zero to the nonzero steady states. The thicker line is a Hopf bifurcation line. The dot marks a double zero eigenvalue.

Hopf bifurcation to stable oscillations and the lower stability limit is a steady bifurcation to stable non-zero steady states.

The pupil eye reflex, the postural control, and the stick-balancing experiments are based on the fact that perceiving a stimulus and initiating an action take about 0.1 s each because of the time electrical impulse needs to travel through nerve cells. By the time a person's hand is trying to grasp a moving object, say a ball, the hand's position is being guided by information that is already 0.2 s old. The ball may have slightly changed its trajectory. Researchers believe the visual-motor system accounts for this delay by predicting an object's trajectory and moving the body in accordance with that prediction, like a quarterback throwing the ball ahead of a running receiver. In [112], the authors asked people to use a computer mouse to track a moving target on a screen. The results confirmed that the hand motion precedes on the average the target motion suggesting one way the brain's visual-motor system compensates for the lag time between perception and action.

4

Bernoulli's equation

Daniel Bernoulli (1700–1782) disclosed the equation used most frequently in engineering hydraulics in 1738 (*Hydrodynamica*). This equation relates the pressure, velocity, and height in the steady motion of an ideal fluid. The modern form of his equation is $v^2/2 + p/\rho + gz = cst$, where v is the velocity at a point, p the pressure, ρ the density, g the acceleration of gravity, and z the height above an arbitrary reference level. The simplest derivation comes from the conservation of energy and Bernoulli himself took an equivalent approach.

The application of Bernoulli's equation leads to simple physical laws. In the classical problem of a hole in the side of a tank, we imagine a streamline beginning at the free surface, where the velocity is zero, and extending into the jet a distance h below. The pressures at the two points are the same: atmospheric. From Bernoulli's equation, we then get $v = \sqrt{2gh}$. Another simple law relates the change of velocities to the change of pressure in a horizontal tube. If the output velocity is zero, the input velocity is $v = \sqrt{2\Delta p/\rho}$ where Δp is defined as the change of pressure.

What happens if Bernoulli flow undergoes delayed feedback? This occurs when a control is acting on the flow either artificially through an external device (see Section 4.3) or naturally through a physiological control (see Section 4.2). Another interesting case appears for reed wind instruments such as the clarinet where the delay comes from the round trip of a pressure wave in the pipe (See Section 4.1).

T. Erneux, *Applied Delay Differential Equations*, Surveys and Tutorials
in the Applied Mathematical Sciences 3, DOI 10.1007/978-0-387-74372-1_4,
© Springer Science+Business Media, LLC 2009

4.1   The clarinet

Sound production in reed wind instruments is the result of self-sustained oscillations. A mechanical oscillator, the reed, acts as a valve that modulates the air flow entering into the mouthpiece. See Figure 4.1. The destabilization of the mechanical element is the result of a complex aeroelastic coupling among the reed, the air flow into the instrument driven by the mouth pressure of the musician, and the resonant acoustic field in the instrument itself. Following McIntyre et al. [149], wind instruments can be described in terms of a lumped model formed by a closed feedback loop operating as a self-sustained oscillator. In their model, the loop is made of two elements, a lumped nonlinear element (the mouthpiece) and a linear passive element (the resonator, that is, the instrument itself). The modeling and measurement of the resonator have been studied extensively by many authors but the nonlinear element and its action have only recently been addressed.

The nonlinear element can be defined by a relationship between the pressure difference across the reed and the volume flow at the inlet of the pipe of the instrument. By modulating the aperture height $H(t)$ between the reed and the mouthpiece, the musician controls the volume velocity $u(t)$ through the reed slit. The latter is related to the pressure difference $\Delta p \equiv p_m - p$ through the Bernoulli equation

$$u = wH\sqrt{\frac{2\Delta p}{\rho}}, \qquad (4.1)$$

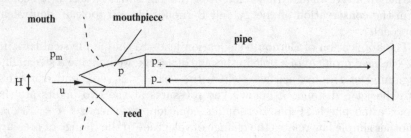

Figure 4.1: Schematic description of the clarinet. The physical model of a clarinet combines a passive resonator (the pipe) and a nonlinear element (the mouthpiece). An aeroelastic instability leads to self-sustained oscillations that produce the sound of the instrument. The reed acts as a valve that modulates the air flow entering into the mouthpiece. $H(t)$ $(0-1\text{ mm})$ is the aperture height and u is the air volume velocity $(100-2000\text{ cm}^3\text{s}^{-1})$. It depends nonlinearly on the pressure difference $\Delta p = p_m - p$ between the mouth pressure p_m $(0-150\text{ hPa})$ and the acoustic pressure $p(t)$ in the mouthpiece. The latter is the sum of the positive and negative plane wave pressures at the input (p_+ and p_-, respectively).

where w is the width of the slit and ρ is the density of air. The reed is assumed to behave as an ideal spring characterized by its stiffness K. $H(t)$ then is a linear function of the pressure difference Δp given by

$$H = H_0(1 - \frac{\Delta p}{p_M}) \qquad (4.2)$$

if $\Delta p \leq p_M \equiv KH_0$ and is zero if $\Delta p > p_M$. H_0 is the opening at rest. Using (4.2), we eliminate H in (4.1) and find u as a function of the normalized pressure difference $\Delta p/p_M$. It is given by

$$u = F(p) \equiv u_0(1 - \frac{\Delta p}{p_M})\sqrt{\frac{\Delta p}{p_M}} \text{ if } \Delta p \leq p_M,$$

$$= 0 \text{ if } \Delta p > p_M. \qquad (4.3)$$

The function has a typical parabolic form starting from $\Delta p = 0$ and ending at $\Delta p = p_M$. The volume velocity parameter u_0 is the only parameter and is defined by

$$u_0 \equiv wH_0\sqrt{\frac{2KH_0}{\rho}}. \qquad (4.4)$$

Its value ranges from 100 to 2000 cm^{3}s^{-1}.

We next need to take into account the effect of the resonator. The acoustic pressure $p(t)$ and the volume velocity $u(t)$ are related to the outgoing and ongoing plane wave pressures p_+ and p_- as $p = p_+ + p_-$ and $u = Z_c^{-1}(p_+ - p_-)$ where $Z_c = \rho c/S$ is the characteristic impedance of the pipe and S is its cross-section area. Equivalently, we may express p_+ and p_- in terms of p and u as

$$p_+ = \frac{1}{2}(p + Z_c u), \qquad (4.5)$$

$$p_- = \frac{1}{2}(p - Z_c u). \qquad (4.6)$$

The outgoing and ongoing wave pressures are related. The simplest way is to assume that

$$p_- = -p_+(t - \tau), \qquad (4.7)$$

where $\tau = 2L/c$ is the round-trip time of a wave at speed c along the pipe of length L.[1] The main weakness of this lossless model is the fact that it allows periodic oscillations for every mouth pressure above the threshold of oscillations and that it shows no extinction phenomenon. Atig et al. [7]

---

[1]More precisely, we assume a delayed delta function for the reflection coefficient of the form $r(t) = -\delta(t - \tau)$. The positive and negative plane-wave pressures are related by the convolution integral as $p_- = r(t) * p_+(t)$, and after inegration, we obtain (4.7). For (4.8), the reflection function is given by $r(t) = -\exp(-2\alpha L)\delta(t - \tau)$.

have recently demonstrated that losses are responsible for this extinction phenomenon. Dalmont et al. [50] take into account losses by assuming that

$$p_- = -\exp(-2\alpha L)p_+(t - \tau), \tag{4.8}$$

where α is defined as a frequency-independent absorption coefficient. Using (4.5) and (4.6), Eq. (4.8) can be rewritten as

$$p - Z_c u = -\exp(-2\alpha L)(p(t - \tau) + Z_c u(t - \tau)). \tag{4.9}$$

This equation is a difference equation from which we may formulate a map relating $(p, u) = (p_n, u_n)$ and $(p, u) = (p_{n-1}, u_{n-1})$. The equation for the map has the form

$$p_n - Z_c u_n = -\exp(-2\alpha L)(p_{n-1} + Z_c u_{n-1}). \tag{4.10}$$

We next rewrite our equations in dimensionless form by introducing x_n and y_n defined by

$$x_n \equiv \frac{Z_c}{P_M} u_n \quad \text{and} \quad y_n \equiv \frac{p_n}{P_M}. \tag{4.11}$$

In terms of x_n and y_n, Eqs. (4.10) and (4.3) become

$$y_n - x_n = -\exp(-2\alpha L)(y_{n-1} + x_{n-1}), \tag{4.12}$$

$$x_n = F(y_n), \tag{4.13}$$

where

$$F(y) \equiv \zeta(1 - \gamma + y)\sqrt{\gamma - y} \quad \text{if } \gamma - y \leq 1,$$
$$= 0 \text{ if } \gamma - y > 1. \tag{4.14}$$

The new bifurcation parameter γ and the fixed parameter ζ are defined by

$$\gamma \equiv \frac{p_m}{p_M} \quad \text{and} \quad \zeta \equiv Z_c u_0/p_M. \tag{4.15}$$

A square-wave periodic solution (period 2τ) satisfying the condition $x_{n+2} = x_n$ and $y_{n+2} = y_n$ is the simplest periodic solution. It is the fundamental regime (first register) and its bifurcation diagram defines the playing range of the instrument. Using Eq. (4.12) for two successive steps leads to the following problem

$$y_2 - x_2 = -\exp(-2\alpha L)(y_1 + x_1), \tag{4.16}$$

$$y_1 - x_1 = -\exp(-2\alpha L)(y_2 + x_2), \tag{4.17}$$

where we have used $y_3 = y_1$ and $x_3 = x_1$ to simplify the left-hand side of Eq. (4.17). Adding and substracting Eqs. (4.16) and (4.17) leads to the following equations

$$y_2 + y_1 = (x_2 + x_1)\tanh(\alpha L), \tag{4.18}$$

$$y_2 - y_1 = \frac{1}{\tanh(\alpha L)}(x_2 - x_1). \tag{4.19}$$

Two types of periodic regimes are possible: the nonbeating reed regime and the beating reed regime for which one of the two states of the reed is held motionless against the lay (say $x_2 = 0$). We first concentrate on the beating reed regime. Assuming $x_2 = 0$, we determine y_2 and y_1 from Eqs. (4.18) and (4.19). We find

$$y_1 = x_1/\tanh(2\alpha L) \quad \text{and} \quad y_2 = -y_1/\cosh(2\alpha L). \qquad (4.20)$$

We then use $x_1 = F(y_1)$ where $F(y)$ is defined in (4.14) and obtain an equation for y_1 only of the form

$$\beta y_1 = (1 - \gamma + y_1)\sqrt{\gamma - y_1} \ (\gamma - y_1 < 1), \qquad (4.21)$$

where

$$\beta = \tanh(2\alpha L)/\zeta. \qquad (4.22)$$

Equation (4.21) is valid provided $x_2 = 0$, that is, provided $\gamma - y_2 > 1$. The critical point satisfying the condition $\gamma - y_2 = 1$ defines the beating reed threshold. Using (4.20), this condition can be rewritten as

$$\gamma = 1 - \frac{y_1}{\cosh(2\alpha L)} \qquad (4.23)$$

and after eliminating γ in Eq. (4.21), we find

$$y_1 = \frac{1}{1 + \sec h(2\alpha L)} \left[1 - \left(\frac{\beta}{1 + \sec h(2\alpha L)} \right)^2 \right]. \qquad (4.24)$$

If losses are weak ($\alpha L \ll 1$ and $\beta \ll 1$), $y_1 \simeq 0.5$ and $\gamma \simeq 0.5$.

Equation (4.21) can be reformulated as a cubic polynomial in y_1. Alternatively, we can express the solution in parametric form. This is suggested by the observation that γ in Eq. (4.21) always appears combined with $\gamma - y_1$. Introducing the parameter z as

$$y_1 = \gamma - z \qquad (4.25)$$

Eq. (4.21) leads to

$$\gamma = z + (1 - z)\sqrt{z}/\beta \qquad (4.26)$$

and then using (4.25)

$$y_1 = (1 - z)\sqrt{z}/\beta. \qquad (4.27)$$

Together with (4.26), (4.27) provides the solution $y_1 = y_1(\gamma)$ by changing z continuously ($0 \leq z \leq 1$). See Figure (4.2). The branch starts at the critical point where $\gamma - y_2 = 1$ (beating reed threshold where $x_2 = 0$). It reaches a limit point at $\gamma = \gamma_{extup}$. From (4.26) and the condition $d\gamma/dz = 0$, we

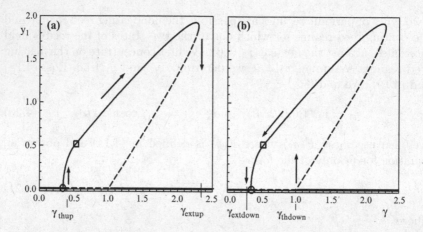

Figure 4.2: Bifurcation diagrams representing the pressure in the mouthpiece $y_1 = p_1/P_M$ as a function of mouth pressure $\gamma = p_m/P_M$. $\beta = 0.2$. Different evolutions are obtained if we increase (a) or decrease (b) γ. Labels O and □ denote the threshold of oscillation and beating reed threshold, respectively.

obtain the following quadratic equation for $\sqrt{z}$: $-3(\sqrt{z})^2 + 2\beta\sqrt{z} + 1 = 0$ leading to $\sqrt{z} = (\beta - \sqrt{\beta^2 + 3})/3$. In the limit β small, we have $z \simeq 1/3$ meaning

$$\gamma_{extup} \simeq \frac{2}{3\sqrt{3}}\beta^{-1}. \tag{4.28}$$

Because $\gamma_{thdown} = 1$, the domain of the hysteresis loop in Figure 4.2 decreases if β is increased.

The nonbeating regime where both x_1 and x_2 are nonzero is more difficult to determine analytically. However, if $\tanh(\alpha L)$ is sufficiently small, there is a simple approximation obtained by setting $\tanh(\alpha L) = 0$ in (4.18) and (4.19). We successively find $x_2 = x_1$, $y_2 = -y_1$, and finally

$$y_1 = -y_2 \simeq \sqrt{(3\gamma - 1)(1 - \gamma)}. \tag{4.29}$$

Equation (4.29) describes the branch of periodic solutions in Figure 4.2 emerging from the oscillation threshold at $\gamma = 1/3$ to the beating reed threshold at $\gamma = 1/2$.

Figures 4.3 show experimental recording investigating the influence of the reed opening. A small reed opening leads to a small playing range. From (4.15) with (4.4), decreasing H_0 implies decreasing ζ and then from (4.22) an increase of β.

When a clarinet is blown with increasing mouth pressure the reed begins to oscillate as the pressure surpasses the threshold of oscillations and stops as the pressure reaches the threshold of extinction.

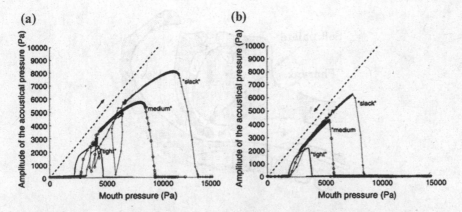

Figure 4.3: Acoustical pressure as a function of the mouth pressure during (a) crescendos and (b) decrescendos. The three curves in each figure correspond to three different embouchures. Tight, medium and slack refers to $H_0 = 0.37, 0.53$, and 0.79 mm, respectively. Reprinted from Atig et al., [7], with permission from Elsevier.

See Figure 4.2. Above this threshold, the reed is held motionless against the lay. We have found that losses in the pipe significantly affect the value of the extinction threshold and, consequently, the maximum power of the instrument.

An essential nonlinearity comes from the Bernoulli equation, $v = \sqrt{2\Delta p/\rho}$, which relates the volume velocity $u = wH(t)v$ of the flow through the reed channel and the pressure difference Δp on both sides of the reed. It is based on the asumption that the kinetic energy of the jet entering the instrument is completely dissipated into turbulence during its expansion in the mouthpiece. Flow-induced vibrations such as the musical sound produced by a reed instrument can be seen as the loss of stability of a mechanical oscillator interacting with a continuous flow. Another documented example of such aeroelastic instability is human snoring.

4.2   Sleep disorders

Sleep disorders affect a measurable segment of the population. Consequently, normal and abnormal functions of the airway are extensively studied from a medical point of view. Two of these disorders, namely snoring and obstructive sleep apnea (OSA) are defined as disorders of the upper airway. The upper airway stretches from the nasal and oral inlets to the larynx (or vocal box), as shown in Figure 4.4. Snoring describes the audible sound generated by a fluttering soft palate. OSA mainly affects the pharyngeal

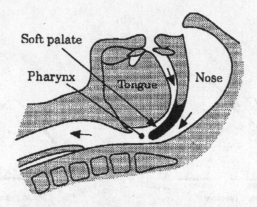

Figure 4.4: Upper airway narrowing is known as a predisposing factor for snoring and obstructive sleep apnea. It occurs in the pharynx near the soft palate. Reprinted from Aurégan and Depollier [9], with permission from Elsevier.

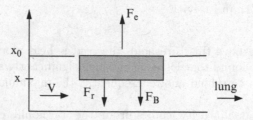

Figure 4.5: Airway model. A collapsible segment is represented as a section of length L, width W, and depth x. It is supported by a spring with elastic constant K and is connected to the rigid upper airway having resistance (R) to flow (V).

region, collapsing it fully or partially, thereby restricting the airflow to the lungs. This flow-induced instability can also be caused by large deflections of the soft palate. Theories of snoring include the early work of Gavriely and Jensen [71], where steady Bernoulli flow with pressure loss due to upper airways resistance enters a spring-loaded channel section. The resulting nonlinear model predicts either flutter oscillations around a neutral wall position, or closure (obstruction) instability, where the channel height decreases to zero as the wall mass overshoots its equilibrium. The model predictions for the closure time course compared favorably with sound recordings of snores. Figure 4.5 gives a schematic representation of the upper airways as described by Gavriely and Jensen [71]. A collapsing segment is represented as a section of length L with a rectangular cross-section of width W and depth x. The moveable wall of the collapsible segment is assumed to have a mass m and to be supported by a linear spring with a Hooke spring constant K. Its neutral position is at $x = x_0$. A constant mean V is assumed in the inspiratory direction. The upper airways, from the collapsible section to the airway opening, are assumed to have a linear flow resistance (R). The intra-airway pressure in the collapsible segment has two components: a pressure drop due to the viscous resistance of the upper airways $(P_r = VR)$ and a Bernoulli effect pressure across the collapsible segment $(P_B = \frac{1}{2}\rho \left(V/(Wx)\right)^2$ where ρ is the gas density). The subatmospheric pressure beneath the movable wall tends to pull it closer to the opposing wall and to reduce x.

The equation of motion of the plate takes into account all forces and is given by

$$m\frac{d^2x}{ds^2} = F_e - F_r - F_B, \qquad (4.30)$$

where s is time. The elastic viscous resistance and Bernoulli forces are defined by

$$F_e \equiv K(x_0 - x), \quad F_r \equiv P_r LW = VRLW,$$

$$F_B \equiv LWP_B = LW\frac{1}{2}\rho \left(\frac{V}{Wx}\right)^2. \qquad (4.31)$$

In this purely mechanical model, the neuromuscular reflex is ignored. Huang and Williams [103] coupled snoring mechanics with neurological responses, which tend to contract (stiffen) pharyngeal muscles (walls). If the elastic forces are insufficient to maintain the stability of the airway, neuromuscular functions become crucial. However, these functions are very much reduced during sleep, and the muscle reflex mechanism may have a time delay of several cycles of oscillations experienced during snoring. Huang and Williams assumed that following a delayed signal from the neural receptors, the dilator muscle opposes the collapsing tendency by increasing the wall stiffness by an amount proportional to the negative pressure at time $s - T$. Huang and Williams [103] formulated a linearized theory and ignored the viscous resistance. Here, we add the delayed rescuing force in Eq. (4.30) and find

$$m\frac{d^2x}{ds^2} = F_e - F_r - F_B + \alpha\left[F_r(s - T) + F_B(s - T)\right], \qquad (4.32)$$

where α is a dimensionless parameter. Its value is unknown but the range of T is available for muscle reflex during wakefulness and sleep. Introducing the dimensionless variables and parameters defined in [71], Eq. (4.32) simplifies as

$$\frac{d^2y}{dt^2} + \delta\frac{dy}{dt} = 1 - y - q(1 - \alpha) - \frac{\mu q^2}{2}\left[\frac{1}{y^2} - \frac{\alpha}{y^2(t - \tau)}\right], \qquad (4.33)$$

where we have added damping and the damping parameter δ is small and positive. The parameter q is proportional to the inspiratory flow V and μ

is independent of V. Using reasonable values of the physical parameters, Gavriely and Jensen [71] estimated the values of q and μ for normal breathing conditions close to 0.3 and 1, respectively. The steady-state solutions satisfy the following equation

$$1 - y - q(1 - \alpha) - \frac{\mu q^2}{2}\frac{1 - \alpha}{y^2} = 0. \qquad (4.34)$$

Taking the derivative of Eq. (4.34) with respect to y indicates that $dq/dy > 0$ $(dq/dy < 0)$ if

$$1 - \frac{\mu q^2}{y^3}(1 - \alpha) < 0 \ (> 0). \qquad (4.35)$$

The bifurcation diagram for two different values of α is shown in Figure 4.6. If $\alpha = 0$, only the upper branch is stable [71]. We examine the linear stability of the steady states if $\alpha \neq 0$. The linearized problem is given by

$$\frac{d^2 u}{dt^2} + \delta\frac{du}{dt} + u - \frac{\mu q^2}{y^3}(u - \alpha u(t - \tau)) = 0. \qquad (4.36)$$

Introducing $u = \exp(\lambda t)$ into Eq. (4.36) leads to the following characteristic equation

$$\lambda^2 + \delta\lambda + 1 - \frac{\mu q^2}{y^3}(1 - \alpha\exp(-\lambda\tau)) = 0. \qquad (4.37)$$

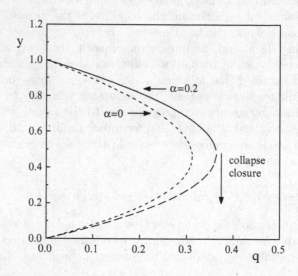

Figure 4.6: Steady-state solutions. The upper and lower branches correspond to stable (center) and unstable (saddle) steady state solutions, respectively. $\mu = 1$ and $\alpha = 0$ or $\alpha = 0.2$. The neuromechanical effect is stabilizing because the limit point above which collapse appears is moved to a larger value of q.

Assuming τ small and expanding the exponential in (4.37), we find

$$\lambda^2 + (\delta - \frac{\mu q^2}{y^3}\alpha\tau)\lambda + 1 - \frac{\mu q^2}{y^3}(1 - \alpha) \simeq 0, \qquad (4.38)$$

where we note a decrease of the damping coefficient multiplying λ. This crucial point was already observed by Huang and Williams [103]. If the damping coefficient becomes negative, flutter occurs and this instability results from the muscle rather than the flow. The critical point where the damping coefficient is zero corresponds to a Hopf bifurcation point given by

$$\tau_H = \frac{\delta y^3}{\mu q^2 \alpha} \qquad (4.39)$$

in the $\tau = O(\delta) \to 0$. The oscillatory instability occurs as soon as $\tau > \tau_H$ which suggests that τ needs to be sufficiently small in order to guarantee stability. But we need to remember that our analysis is valid if $\tau = O(\delta)$. If τ is arbitrary, we need to determine the Hopf stability boundary in a different way. Introducing $\lambda = i\omega$ into Eq. (4.37) leads to the two Hopf conditions for the frequency ω and amplitude q^2/y^3:

$$\omega^2 - 1 + \frac{\mu q^2}{y^3}(1 - \alpha \cos(\omega\tau)) = 0, \qquad (4.40)$$

$$\delta\omega - \frac{\mu q^2}{y^3}\alpha \sin(\omega\tau) = 0. \qquad (4.41)$$

The numerical solution of the first Hopf boundary is shown in Figure 4.7. The lowest broken line is given by (4.39). The upper broken line is given by

$$\tau = \frac{\pi}{\sqrt{1 - \mu q^2(1 + \alpha)y^{-3}}} \qquad (4.42)$$

4.3   Cascaded control of a liquid level system

A classical liquid level control system is perhaps the simplest example of a Bernoulli problem where the delay may play a major role [31]. In Figure 4.8, the control variable is the inlet flow rate into the tank. The outlet flow is assumed to be turbulent such that a nonlinear Bernoulli equation relates the liquid level to the outlet flow rate. The mass balance equation is of the form

$$\frac{dx}{ds} = c_1(u_s + u) - c_2\sqrt{h_s + x} \qquad (4.43)$$

where $x \equiv h - h_s$ is the liquid level deviation from its steady-state value and u_s is the steady state inlet flow rate satisfying the steady-state condition

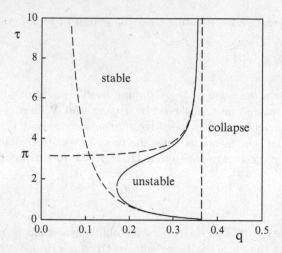

Figure 4.7: First Hopf bifurcation line (full line). $\mu = 1, \alpha = 0.2$, and $\delta = 0.01$. The broken lines are the asymptotic approximations for $\delta \to 0$.

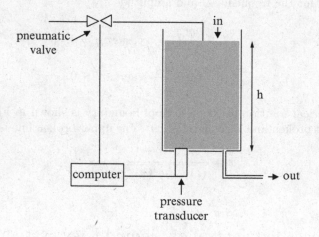

Figure 4.8: Experimental set-up.

$c_1 u_s = c_2 \sqrt{h_s}$. The coefficient c_1 is the inverse tank cross-sectional area and c_2 is related to the valve coefficient of the outlet orifice. Introducing the new variables $y \equiv c_2^2 x$, $t \equiv c_2^2 s$, and the new parameters $v \equiv c_1 u$ and $c \equiv c_1 u_s$, Eq. (4.43) simplifies as

$$\frac{dy}{dt} = c + v - \sqrt{y + c^2}. \qquad (4.44)$$

We next introduce a cascaded control scheme. The inner loop consists of a PI (Proportional-Integral) controller of the form

$$v = -k_1(y - y_{set}) - k_2 \int_0^t (y - y_{set})dt, \qquad (4.45)$$

where k_1 and k_2 are defined the proportional and integral gains, respectively. The outer loop is controlled by a proportional controller with gain k and delay τ of the form

$$y_{set} = ky(t - \tau). \qquad (4.46)$$

The integrodifferential delay equation (4.44)–(4.46) can be rewritten into a set of DDEs by introducing the new variable

$$z \equiv -k_2 \int_0^t (y - y_{set})dt \qquad (4.47)$$

into (4.45) and by differentiating (4.47). We then obtain

$$\frac{dy}{dt} = c - k_1(y + ky(t - \tau)) + z - \sqrt{y + c^2}, \qquad (4.48)$$

$$\frac{dz}{dt} = -k_2(y + ky(t - \tau)). \qquad (4.49)$$

From the linearized equations for $y = 0$, we determine the characteristic equation for the growth rate σ. It is given by

$$(1 + k\exp(-\sigma\tau))(\sigma k_1 + k_2) + \sigma^2 + \sigma/2c = 0. \qquad (4.50)$$

The Hopf bifurcation boundaries are obtained by inserting $\sigma = i\omega$ into (4.50) and by separating the real and imaginary parts. We obtain the following two conditions for k and τ,

$$k\cos(\omega\tau) = A \equiv \frac{\omega^2(k_2 - k_1/2c)}{k_2^2 + k_1^2\omega^2} - 1, \qquad (4.51)$$

$$k\sin(\omega\tau) = B \equiv \frac{\omega(k_1\omega^2 + k_2/2c)}{k_2^2 + k_1^2\omega^2}. \qquad (4.52)$$

The possible solutions are obtained by continuously changing ω from a large negative value to a large positive value and by computing $k = k(\omega)$ from $k = \sqrt{A^2 + B^2}$. Knowing $k = k(\omega)$, we then determine $\tau = \tau(\omega)$ using Eq. (4.51). The possible solution branches are

$$\tau = \tau_1(n) \equiv (\arccos(A/k) + 2n\pi)/\omega, \qquad (4.53)$$

$$\tau = \tau_2(m) \equiv (-\arccos(A/k) + 2m\pi)/\omega, \qquad (4.54)$$

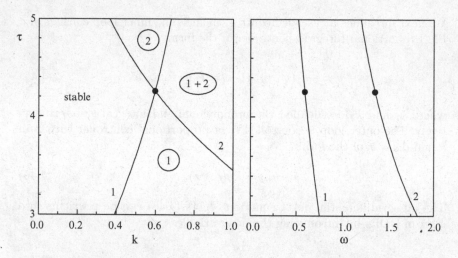

Figure 4.9: Hopf stability boundaries. Left: Stability diagram in terms of τ versus k. The Hopf bifurcation lines 1 and 2 correspond to $n = 0$ and $m = 1$, respectively. Regions 1, 2, and 1+2 represent the domains of oscillations with frequency ω_1, frequency frequency ω_2, and quasiperiodic oscillations with frequencies ω_1 and ω_2, respectively. Right: Hopf bifurcation frequencies associated with the Hopf lines 1 and 2. The values of the fixed parameters are $k_1 = -0.8$, $k_2 = 1$, and $c = 1/2$.

where $n = 0, 1, \ldots$ and $m = 1, 2, \ldots$ are integers. Figure 4.9 represents the Hopf bifurcation lines in the (k, τ) parameter space for $n = 0$ (line 1)and $m = 1$ (line 2). The two lines cross at the double Hopf point located at

$$(k^*, \tau^*) = (0.5969, 4.2565). \qquad (4.55)$$

This Hopf bifurcation corresponds to two pair of imaginary eigenvalues or two different frequencies given by

$$\omega_1^* = 0.5918 \quad \text{and} \quad \omega_2^* = 1.3556. \qquad (4.56)$$

The experiments used a cylindrical clear plastic tank with diameter of 20 cm and height of approximatively 110 cm. The inlet at the top of the tank was fed through a one inch pneumatic control valve. The height was inferred by measuring the hydrostatic pressure at the bottom of the tank with a Teledyne pressure transducer. The signal from the pressure transducer was sent to the computer which in turn output the appropriate delayed signal to the control valve (Figure 4.8). Figure 4.10 shows the experimental Hopf bifurcation points in the (k, τ) parameter space. These points were determined at fixed delay τ by increasing k until the system was destabilized. The gain k was then decreased to check for any hysteresis. In all these experiments, the stability boundary did not exhibit any hysteresis suggesting

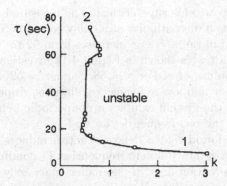

Figure 4.10: Experimental stability boundary. A crossing of two distinct Hopf bifurcation lines with Hopf frequencies ω_1 and ω_2 occurs near $\tau = 60$ s. As τ increased from 52 to 58 s, the Hopf frequency jumps from a low to a higher value. Reprinted from Boe and Chang [31], with permission from Elsevier.

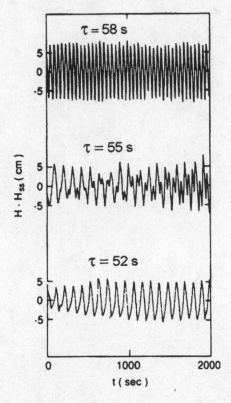

Figure 4.11: Experimental time traces of the liquid level oscillations for k slightly above its Hopf bifurcation value. Reprinted from Boe and Chang [31], with permission from Elsevier.

that the Hopf bifurcation is supercritical. The measured Hopf bifurcation frequency ω exhibits a discontinuity at a delay close to 55 s. It jumps from 0.059 rad s^{-1} to 0.130 rad s^{-1} as τ increases from 52 to 58 s.

The experimental traces shown in Figure 4.11 investigate the transition from high to low values of τ. For $\tau = 58$ s and $\tau = 52$ s, the system exhibits high-frequency and low-frequency oscillations, respectively. Between these two regimes, the system exhibits a nonperiodic behavior suggesting quasiperiodic with the two frequencies ω_1 and ω_2.

In summary, we found that a delayed control induces several Hopf bifurcations characterized by different frequencies. Depending on the parameters, the Hopf bifurcation lines in parameter space may cross generating double Hopf bifurcation points. Near those points, the dynamical response of the system can be rich and includes jumps between oscillations at different frequencies or quasiperiodic oscillations. The bifurcation analysis at and near a double Hopf bifurcation is analyzed in detail in Chapter 7.

5
Chemistry

Chemists have been making a serious study of the rates of reactions since the middle of the 19th century. Often they found that the rates were proportional to the concentrations of the substances that were reacting together. In 1892, however, Adrian John Brown, Professor of Malting and Brewing at Birmingham found that the rate of fermentation of sucrose in the presence of yeast seemed to be independent of the amount of sucrose present [35]. Ten years later, he explained his results by assuming that the invertase molecules present in yeast formed an addition complex with sucrose [36]. This was the first time that the existence of an enzyme–substrate complex was deduced from the kinetics of an enzyme reaction.[1] Brown thought that the enzyme–substrate complex needed a fixed lifetime to decay rather than decaying by the usual first-order rate law [36]. Although the idea was not popular at the time of Brown, there is evidence today that we need to take into account the duration of conformational changes that take place at the molecular level during the catalytic cycle of the monomer [2]. Moreover, there has been a renewed interest for DDEs since the end of the 1990s and, in particular, for model reduction [202, 54, 94]. Consider the following sequence of reactions modeling the enzymatic transformation of a substrate S into a product P

---

[1]While writing his paper, Brown became aware of the work by Victor Henri [89] who wrote down a mathematical expression for the rate of the enzymatic reaction that we now know as the Michaelis–Menten reaction [160].

T. Erneux, *Applied Delay Differential Equations*, Surveys and Tutorials in the Applied Mathematical Sciences 3, DOI 10.1007/978-0-387-74372-1_5, © Springer Science+Business Media, LLC 2009

$$S + E \xrightarrow{k_0} C_1 \xrightarrow{k_1} ... C_n \xrightarrow{k_n} E + P. \tag{5.1}$$

The question is whether we may describe this cascade of reactions (5.1) by the following reaction

$$S + E \xrightarrow{delay} P + E \tag{5.2}$$

if the number of intermediates N is sufficiently large. In 2004, Hinch and Schnell [94] showed that the rate of P is directly related to the decrease of S as

$$\frac{dS}{dt} = -R(t) \quad \text{and} \quad \frac{dP}{dt} \simeq R(t - \tau) \tag{5.3}$$

provided that $\tau \equiv \sum_{i=2}^{N} k_i^{-1} \ll k_1^{-1}$. The inclusion of the delay simplifies the mathematical description of the kinetic model because we don't need to know all its details such as the k_j.

In a different setting, delay cannot be ignored in industrial applications where recycling of unreacted reagents reduces the cost of the reaction. The output stream of a continuously stirred tank reactor (CSTR) is sent through a separation process. Then the unreacted reagents are returned into the reactor by traveling through pipes. This process requires a finite amount of time and will introduce a delay into the model because both the concentration of the reagents and the temperature in the reactor depend on some time in the past. In practice (industry), however, it is quite common to ignore the recycle delay and to use standard ODE models [29, 234, 185]. In the 1980s, chemical engineers had the idea to stabilize unstable states in bistable chemical systems by using feedback controls. On the practical side, the deliberate operation of chemical reactors near or at an unstable steady state could allow for a higher yield and/or selectivity. In 1984, Zimmermann et al. in [247] used a delayed feedback loop to stabilize the unstable branch of steady states of an illuminated thermochemical reaction. Photochemical reactions offer a convenient opportunity to introduce delays because it is relatively easy to vary the intensity of illumination in response to the value of some measured property at an earlier time.

Although Zimmermann et al. [247] successively stabilized unstable steady states using a delayed feedback with a small delay, the delay is not the mechanism responsible for the stabilization of the unstable steady states but rather a clever feedback design. Nevertheless, the authors discovered that limit-cycle oscillations appear if the delay is sufficiently large and that these oscillations result from a Hopf bifurcation mechanism. In the first part of this chapter, we review their work on the illuminated thermochemical reaction and then consider a different bistable system which was investigated by Laplante in 1989. In the second part of this chapter, we examine strongly oscillating chemical reactions subject to a weak delayed feedback.

5.1   Illuminated thermochemical reaction

When gaseous $S_2O_6F_2$ is illuminated at 488 nm, only the product SO_3F absorbs the light, which is converted to heat. As a result of the heating, the equilibrium shifts to the right, causing more SO_3F to be produced, and increasing further the amount of light absorbed. The reaction thus exhibits a form of autocatalysis. If the incident light intensity is slowly varied, the system shows hysteresis between a high-monomer and a low-monomer steady state [246]. We next describe the formulation of the rate equations. The gaseous reaction mixture

$$S_2O_6F_2 \rightleftharpoons 2SO_3F$$

at a total fixed pressure p_0, is contained in a cell of length l, with the walls of the cell held at a temperature T_b. A laser beam of radius r and power Φ_0 is incident on a small region of the cell at a wavelength absorbed only by SO_3F. The system is defined as the volume of the cell illuminated by the beam. The absorbed light causes the temperature rise which drives the thermochemical system out of equilibrium. Energy is lost through the boundaries by heat transfer. The state of the system is described by the concentration of SO_3F and by the temperature [X (in mol cm^{-3}) and T (in K), respectively]. Their kinetic equations are given by

$$\frac{dX}{dt} = k_2 \left[\frac{\alpha}{T} \exp(-\Delta H/RT)(\frac{p_0}{RT} - X) - X^2 \right], \qquad (5.4)$$

$$C\frac{dT}{dt} = A(X)\Phi_0 - \beta(T - T_0) - \lambda\frac{dX}{dt}, \qquad (5.5)$$

where $A(X) \equiv 1 - \exp(-\varepsilon X l)$. Equation (5.4) is the chemical kinetic relation for the reaction including the pressure constraint. Equation (5.5) is the thermal energy balance equation. The terms in the right-hand side of Eq. (5.5) represent, respectively, the power input due to absorption according to Beer's law, the heat transfer rate from the system, and the rate of enthalpy change due to chemical reaction. A is the total absorption of light by the reaction mixture that can be measured directly. Φ_0 is the light power measured in W.

5.1.1   Reformulation

Introducing the dimensionless variables $x \equiv X/X_0$, $\theta \equiv T/T_0$, and $s \equiv t/t_0$ where

$$X_0 \equiv \frac{1}{\varepsilon l}, \quad T_0 \equiv T_b \quad \text{and} \quad t_0 \equiv \frac{1}{k_2 X_0}, \qquad (5.6)$$

we rewrite Eqs. (5.4) and (5.5) as

$$\frac{dx}{ds} = \frac{1 - a\theta x}{\theta^2} \exp\left[E(\frac{1}{\theta^*} - \frac{1}{\theta})\right] - x^2, \tag{5.7}$$

$$c\frac{d\theta}{ds} = A(x)\Phi_0 - b(\theta - 1) - \Lambda\frac{dx}{ds}, \tag{5.8}$$

where

$$A(x) \equiv 1 - \exp(-x). \tag{5.9}$$

The new parameters E, θ^*, a, b, c, and Λ are defined by

$$E \equiv \frac{\Delta H}{RT_0}, \quad \exp(E/\theta^*) \equiv \frac{\alpha p_0}{RX_0^2 T_0^2}, \quad a \equiv \frac{RT_0 X_0}{p_0},$$

$$b \equiv \beta T_0, \quad c \equiv CT_0 k_2 X_0, \quad \text{and} \quad \Lambda \equiv \lambda k_2 X_0^2. \tag{5.10}$$

Using the values of the parameters documented in [247], we find $X_0 = 5 \times 10^{-7}$mol cm^{-3}, $T_0 = 333$ K, and $t_0 = (1/750)$. The values of the parameters are listed in the following table.

parameters	values
E	33.33
θ^*	1.29
a	4.26×10^{-2}
c	0.51 W
b	0.81 W
Λ	3.38×10^{-2}

We note that a and Λ are small. Neglecting them, Eqs. (5.7) and (5.8) simplify as

$$x' = \frac{1}{\theta^2} \exp\left[E(\frac{1}{\theta^*} - \frac{1}{\theta})\right] - x^2, \tag{5.11}$$

$$c\theta' = A\Phi_0 - b(\theta - 1), \tag{5.12}$$

where A is given by (5.9). In parametric form, the steady state solution is given by

$$x = \frac{1}{\theta} \exp\left[\frac{E}{2}(\frac{1}{\theta^*} - \frac{1}{\theta})\right], \tag{5.13}$$

$$\Phi_0 = \frac{b(\theta - 1)}{1 - \exp(-x)}. \tag{5.14}$$

The analytical steady state solution is compared to the exact numerical solution in Figure 5.1.

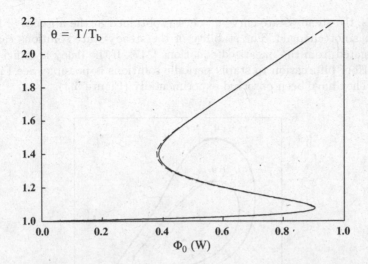

Figure 5.1: Full and broken lines represent the exact steady state and its approximation, respectively. The upper and lower branches are stable and the middle branch is unstable. There is no Hopf bifurcation.

5.1.2 Feedback

Zimmermann et al. [247] introduce a simple feedback in the light power Φ_0. The light output of the feedback loop is given by

$$\Phi_0(t + \tau) = C_1 + C_2(1 - A(t))\Phi_0(t), \qquad (5.15)$$

where C_1 and C_2 are two new parameters. The incident power Φ_0 is no longer a constraint parameter as it is the system without feedback but is now a dependent variable coupled to the concentration X and the temperature T. In terms of the dimensionless time variable s, Eq. (5.15) for Φ_0 becomes

$$\Phi_0(s + s_d) = C_1 + C_2(1 - A(s))\Phi_0(s), \qquad (5.16)$$

where $s_d \equiv \tau/t_0 = 0.75\,ms^{-1} \times \tau$ and τ is measured in ms. The steady-state solutions of Eqs. (5.11), (5.12), and (5.16) now satisfy the three conditions

$$\frac{1}{\theta^2} \exp\left[E\left(\frac{1}{\theta^*} - \frac{1}{\theta}\right)\right] - x^2 = 0, \qquad (5.17)$$

$$A\Phi_0 - b(\theta - 1) = 0, \qquad (5.18)$$

$$-\Phi_0 + C_1 + C_2(1 - A)\Phi_0 = 0. \qquad (5.19)$$

If C_1 is our new control parameter, the steady-state solution admits the parametric solution (5.13) and (5.14), and

$$C_1 = -C_2(1 - A)\Phi_0 + \Phi_0. \qquad (5.20)$$

Thus, the steady-state curve $\theta = \theta(\Phi_0)$ obtained in the absence of feedback has not changed. The stability of the steady-state solutions can be investigated from the linearized equations [247]. If the delay is sufficiently large, Hopf bifurcation to stable periodic solutions is possible. See Figure (5.2). They have been observed experimentally (Figure 5.3).

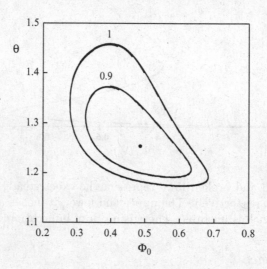

Figure 5.2: Emergence of limit-cycle oscillations as $s_d > 0.82$ ($\tau \simeq 1.09$ ms). Two limit-cycles have been obtained numerically from Eqs. (5.11), (5.12), and (5.16) for $s_d = 0.9$ and $s_d = 1$, respectively. The values of the parameters E, c, b, θ^* are listed in the previous table. $C_1 = 0.2625$ and $C_2 = 0.8$.

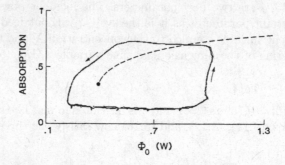

Figure 5.3: Experimental limit-cycle in the (A, Φ_0) plane. $C_2 = 1.5$ and $\tau = 40$ ms. The dashed curve is the stable upper absorption branch obtained by slowly changing C_1 from high to low values. At the point marked by a dot, the steady state becomes unstable and the system spirals out to a limit-cycle (solid closed curve). Three revolutions have been recorded and the arrows indicate the direction of motion of the cycle reprinted with permission from Zimmermann et al. [247] Copyright 1984 American Institute of physics.

5.2 The bistable iodate–arsenous acid reaction

Under excess arsenous acid conditions, the iodate–arsenous acid reaction in a continuously stirred tank reactor can be conveniently described by a first-order nonlinear equation [137],

$$\frac{dY}{dT} = (k_1 + k_2 Y)(X_0 + Y_0 - Y)Y + K(Y_0 - Y), \qquad (5.21)$$

where Y $(= [I^-])$ denotes the concentration of iodide and T is time. X_0 $(= [IO_3^-]_0)$ and Y_0 $(= [I^-]_0)$ represent the reactant stream concentrations of iodate and iodide, respectively. k_1 and k_2 are defined as $k_1 = k_{B1}[H^+]^2 k_1 = k_{B1}[H^+]^2$ where k_{B1} and k_{B2} are kinetic constants. K is the reciprocal residence time and is proportional to the total flow rate. Equation (5.21) can be simplified if we introduce the following dimensionless variables

$$y = \frac{Y}{X_0}, \quad t = k_1 X_0 T, \quad \text{and} \quad k = \frac{K}{k_1 X_0}. \qquad (5.22)$$

In terms of (5.22), Eq. (5.21) becomes

$$\frac{dy}{dt} = y(1 + \alpha y)(1 + \delta - y) + k(\delta - y). \qquad (5.23)$$

The dimensionless parameters α, δ, and k are defined by

$$\alpha = \frac{k_2 X_0}{k_1} = \frac{k_{B2} X_0}{k_{B1}}, \quad \delta = \frac{Y_0}{X_0}, \quad \text{and} \quad k = \frac{K}{k_1 X_0}. \qquad (5.24)$$

The values of the original parameters and α and δ are given in the following table.

k_{B1}	$4.5 \times 10^3 \ M^{-1} s^{-1}$
k_{B2}	$4.5 \times 10^8 \ M^{-1} s^{-1}$
$[H^+]$	$1{:}097 \times 10^{-2} \ (pH = 1.96)$
X_0	$7.1 \times 10^{-4} \ M$
Y_0	$3.2 \times 10^{-5} \ M$
α	71
δ	0.045

The steady state solution $y = y(k)$ is given by (in implicit form)

$$k = \frac{y(1 + \alpha y)(1 + \delta - y)}{y - \delta} \qquad (5.25)$$

Laplante [137] proposed to determine the unstable steady states by controlling the flow rate at the successive times $t = t_n \equiv n t_0$ $(n = 0, 1, 2, ...)$. A simpler form of his algorithm is given by

$$k_{n+1} = k_n + c(y_{n+1} - y_n). \qquad (5.26)$$

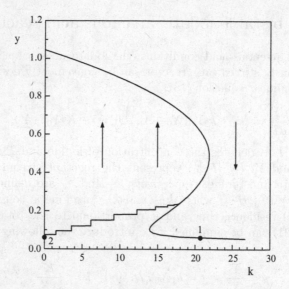

Figure 5.4: Stabilization of an unstable steady-state. The system is first resting at the stable steady state $(20.72, 0.06)$ (Point 1). k is quickly changed to zero (Point 2). The value of k is then sequentially changed following the algorithm (5.26). $c = 100$, $t_0 = 0.04$.

Figure 5.4 shows a numerical simulation leading to the stabilization of an unstable steady state. The first four steps are documented in the following table.

n	t_n	y_n	k_n
0	0.	0.060	0
1	0.04	0.075	1.52
2	0.08	0.096	3.57
3	0.12	0.122	6.18

Proceeding in this way, Laplante [137] determined the branch of unstable saddle-node steady states (see Figure 5.5). The success of this control strategy is essentially the result of the two-dimensional phase-plane (y, k) with restriction of the growth of $y(t)$ (arrows in Figure 5.4). Although several delayed control strategies have successfully stabilized unstable foci near Hopf bifurcation points, they seem unable to stabilize saddle-node steady states [126, 102]. If Laplante were to redo his experiment today, he would presumably use the adaptive controller used by his colleague J. L. Hudson [195]. To stabilize an unstable steady state, the parameter k is perturbed by an adaptive feeback

$$k = k_0 + c_1(w - y), \qquad (5.27)$$

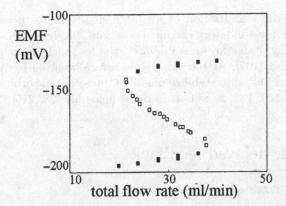

Figure 5.5: Iodate–arsenous acid bistability showing the experimental stable steady states (solid symbol) and unstable states (open symbol). An iodide-selective electrode was used to monitor the iodide concentration. The figure repesents the iodide electrode potential as a function of the total flow rate (proportional to k). Reprinted with permission from Laplante [137] Copyright 1989. America Chemical Society.

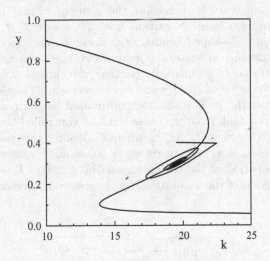

Figure 5.6: Stabilization of the unstable steady state located at $(k, y) = (19.55, 0.3)$ using an adaptive feeddback Initial conditions are $y(0) = 0.4$ and $w(0) = 0.3$. The values of the control parameters are $c_1 = -30$ and $c_2 = 1$.

where w is a dynamical variable of the controller that satisfies

$$w' = c_2(w - y). \tag{5.28}$$

Figure 5.6 shows a numerical example.

Interest in the stabilization of unstable steady states persisted in the 1990s but now for oscillatory chemical reactions [96, 182, 183]. Because the unstable steady states are unstable foci, the feedback delay will have an significant effect [102]. At the same time several delayed feedback control techniques were tested to stabilize unstable time-periodic orbits in chemical systems [209, 140, 182] as well as in other fields [27, 28, 215].

5.3   Weak delayed feedback

5.3.1   Experiments

In 2003, Beta et al. [24] investigated an oscillatory surface chemical reaction (CO oxidation on platinum) and studied the effect of a delayed feedback by controlling the partial pressure of one of the reactants. The control was of the form $p = p_0 + \alpha(I - I(t - \tau))$ where p and I denote the pressure of CO and the integral intensity of a photoemission electron microscope image, respectively (see Figure 5.7). The delay τ is of the same order of magnitude as the period T of the homogeneous limit-cycle oscillations ($T = 2 - 10$ s). By progressively increasing the value of τ, they observed that the period exhibits a jump transition from $T > \tau$ to $T < \tau$ suggesting the possibility of a Z-shaped branch of periodic regimes (see Figure 5.8). However, the existence of two stable regimes for the same value of τ could not be demonstrated experimentally because of technical difficulties. In an earlier study, Weiner et al. [241] were more successful. They examined the effect of delay on the oscillations of the minimal bromate oscillator in a continuous stirred tank reactor. These authors controlled the flow rate as $k = k_0 \left[1 + \beta(C(t - \tau) - C_{av})C_{av}^{-1}\right]$ where C denotes the concentration of ceric ions Ce^{4+}, C_{av} is a constant reference value, and τ ranges from zero to three times the period of the oscillations without delay ($T \sim 10^2$ s). They recorded the period of the oscillations by progressively increasing and then

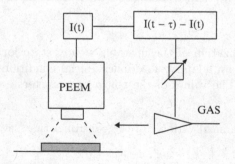

Figure 5.7: PEEM: photoemission electron microscope

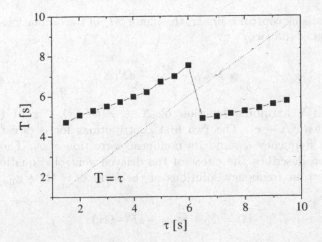

Figure 5.8: Period of the homogeneous oscillations versus the delay reprinted Fig. 3 with permission from Beta et al. [24]. Copyright 2003 by the American physical Society.

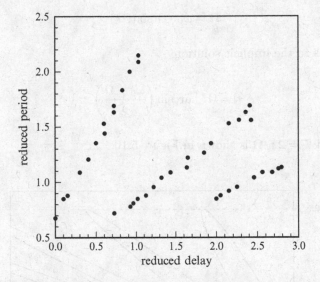

Figure 5.9: Dots: experimental data. Data taken from Figure 3 with permission from Weiner et al. [241]. Copyright 1989. American chemical society.

decreasing τ and found three successive regions where low- and large-period oscillations may coexist (see Figure 5.9). Although Beta et al. [24] could not obtain a similar bifurcation diagram, they provided a first analytical explanation of the phenomena. Close to a supercritical Hopf bifurcation ($\varepsilon^2 = \lambda - \lambda_c \ll 1$) and in limit of weak feedback ($\delta = O(\varepsilon^3)$), only the phase of the nearly harmonic oscillations is affected by the delay. If the delayed feedback control is linear, the phase $\phi(t)$ of the oscillations satisfies an equation of the form

$$\phi' = \omega_0 + \varepsilon^2 \omega_2 + \delta F(\Delta), \qquad (5.29)$$

where F is a harmonic function of $\Delta \equiv \phi(t - \tau) - \phi$ of the form $F = a + b\sin(\Delta + c)$. The two first contributions for ϕ' are the Hopf bifurcation frequency ω_0 and its nonlinear correction $\varepsilon^2 \omega_2$. The small δ contribution describes the effect of the delayed control. Equation (5.29) admits constant frequency solutions of the form $\phi = \Omega t + \phi_0$ where Ω satisfies

$$\Omega = \omega_0 + \varepsilon^2 \omega_2 + \delta F(-\Omega\tau). \qquad (5.30)$$

The solution $\tau = \tau(\Omega)$ can be analyzed by inverting the function F. We illustrate this analysis by considering the case $\omega_2 = a = b - 1 = c = 0$. Equation (5.30) then reduces to

$$\Omega = \omega_0 - \delta \sin(\Omega\tau) \qquad (5.31)$$

which leads to the implicit solution

$$\tau = \Omega^{-1} \arcsin\left(\frac{\omega_0 - \Omega}{\delta}\right). \qquad (5.32)$$

The period $T = 2\pi/\Omega$ is shown in Figure 5.10.

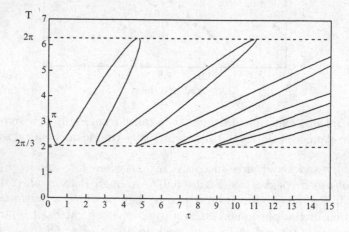

Figure 5.10: Period of the oscillations: $\omega_0 = 2$ and $\delta = 1$.

5.3.2   A piecewise linear oscillator

The theory leading to the phase equation (5.29) is limited to the near vicinity of a Hopf bifurcation and the requirement of very weak feedback. But most of our experimentally studied chemical oscillators exhibit strongly pulsating relaxation oscillations and we wish to examine the effect of delayed feedback on such an oscillator. To this end, we consider a minimal two-variable model described by the following equations,

$$\varepsilon x' = y - f(x), \tag{5.33}$$
$$y' = -x + \delta(x(t - \tau) - x), \tag{5.34}$$

where $\varepsilon \ll 1$ and $f(x)$ is a piecewise linear function of x given by

$$
\begin{aligned}
f(x) &= -x \ (|x| \le 1), \\
&= x - 2 \ (x > 1), \\
&= x + 2 \ (x < 1).
\end{aligned}
\tag{5.35}
$$

The model mimics two variable kinetic models of chemical oscillators exhibiting relaxation oscillations.[2] If $\delta = 0$, Eqs. (5.33) and (5.34) reduces to

$$\varepsilon x' = y - f(x), \quad y' = -x, \tag{5.36}$$

and admits a limit-cycle solution in the phase-plane (x, y) (see Figure 5.11). The oscillations consist of slow evolutions connected by fast transition layers (see Figure 5.12). If $\varepsilon \to 0$, these oscillations approach a discontinuous limit satisfying

$$y_0 = f(x_0) \quad \text{and} \quad y_0' = -x_0. \tag{5.37}$$

Using (5.35), the solution is easily obtained as

$$x_0 = 3\exp(-t), \quad y_0 = x_0 - 2 \ (0 < t \le t_0 = \ln(3)),$$
$$x_0 = -3\exp(-(t - t_0)), \quad y_0 = x_0 + 2 \ (t_0 < t \le T_0 = 2t_0). \tag{5.38}$$

This analytical solution is useful in our analysis of the DDE problem.

If $0 < \delta \ll 1$ we note that the amplitude of the oscillations does not change very much but the period T as a function of the delay admits an interesting behavior. See Figure 5.13. The period varies between two

---

[2]The relaxation oscillations of the Belousov–Zhabotinskii reaction is analyzed in a two-variable phase plane in [172] p 268 and [65] p 149; the two-variable model for the chlorine diode–iodine–malonic acid is described in [219] p 256.

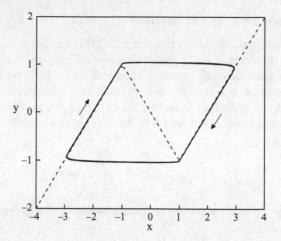

Figure 5.11: Limit-cycle solution for $\delta = 0$ and $\varepsilon = 10^{-2}$. The broken line represents the function $y = f(x)$.

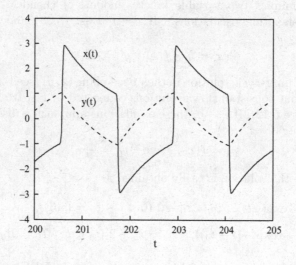

Figure 5.12: Relaxation oscillations for $\delta = 0$ and $\varepsilon = 10^{-2}$. The period of the oscillations is $T_1 = 2.32$.

extrema and exhibits bistability if τ is sufficiently large. The two extrema of the period can be explained as follows. First, we seek a particular solution satisfying

$$x(t - \tau) = x(t). \tag{5.39}$$

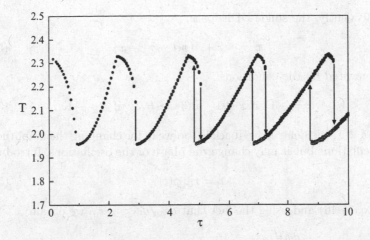

Figure 5.13: Progressive emergence of bistable cycles for the period T as a function of the delay τ. The figure has been obtained by progressevely increasing and then decreasing τ. $\varepsilon = 10^{-2}$ and $\delta = 0.1$.

Inserting (5.39) into Eqs. (5.33) and (5.34), we obtain (5.36) which admits a limit cycle solution of period T_1 (note: $T_1 \to T_0$ as $\varepsilon \to 0$). We then conclude that (5.39) is satisfied if

$$\tau = nT_1 \tag{5.40}$$

for $n = 0, 1, 2...$ Second, we seek another particular solution satisfying

$$x(t - \tau) = -x(t). \tag{5.41}$$

Inserting (5.41) into Eqs. (5.33) and (5.34), we obtain

$$\varepsilon x' = y - f(x), \quad y' = -x(1 + 2\delta). \tag{5.42}$$

These equations admit a limit-cycle of period $T_2 < T_1$ (note: $T_2 \to T_0/(1 + 2\delta)$ as $\varepsilon \to 0$). Because $x(t - T_2/2) = -x(t)$, we conclude that (5.41) is satisfied if

$$\tau = (1 + 2n)\frac{T_2}{2} \tag{5.43}$$

for $n = 0, 1, ...$

In order to demonstrate the bistability phenomenon, we now apply a technique developed by Grasman [79]. In the limit $\varepsilon \to 0$, Eqs. (5.33) and (5.34) reduce to

$$0 = y - f(x), \tag{5.44}$$

$$y' = -x + \delta(x(t - \tau) - x), \tag{5.45}$$

or equivalently, the single DDE for x,

$$x' = -x + \delta(x(t-\tau) - x) \qquad (5.46)$$

supplemented by the conditions

$$x = -3 \quad \text{if } x < 1 \quad \text{and} \quad x = 3 \quad \text{if } x > -1. \qquad (5.47)$$

If $0 < \delta \ll 1$, the delayed feedback is too weak for changing the amplitude of the oscillations but it may change the phase of the oscillations. Introducing

$$x = x_0(\phi(t)) \qquad (5.48)$$

into Eq. (5.46) and using the fact that $dx_0/d\phi = -x_0$, we obtain

$$\frac{d\phi}{dt}\frac{dx_0}{d\phi} = -x_0 + \delta(x_0(\phi(t-\tau)) - x_0),$$
$$\frac{d\phi}{dt} = 1 + \delta - \delta\frac{x_0(\phi(t-\tau))}{x_0}. \qquad (5.49)$$

We now seek a solution of Eq. (5.49) of the form $\phi = \phi_0(t,s) + \delta\phi_1(t,s) + ...$ where $s \equiv \delta t$ is defined as a slow time variable. The leading problem, $\phi_{0t} = 1$, admits the solution

$$\phi_0 = t + \psi(s), \qquad (5.50)$$

where $\psi(s)$ is unknown. The next problem for ϕ_1 then reduces to

$$\phi_{1t} = -\psi_s + a - \frac{x_0(t + \psi(s - \delta\tau) - \tau)}{x_0(t + \psi(s))}, \qquad (5.51)$$

where we assume $\delta\tau = O(1)$. We wish that ϕ_1 remains bounded with respect to the fast time t. This implies the solvability condition

$$\frac{d\psi}{ds} = a - F(\Delta), \qquad (5.52)$$

where

$$F(\Delta) \equiv \frac{1}{T}\int_0^T \frac{x_0(\zeta + \Delta)}{x_0(\zeta)}d\zeta \qquad (5.53)$$

and $\Delta = \psi(s + \delta\tau) - \psi - \tau$ is assumed constant in the ζ integral. We next seek a solution of (5.52) of the form

$$\psi = \sigma s + \psi_0 \qquad (5.54)$$

and compute $F(\Delta)$ where Δ reduces to

$$\Delta = -(1 + \delta\sigma)\tau \qquad (5.55)$$

and is assumed negative. We find

$$F = \frac{\exp(-\Delta)}{t_0} \left[t_0 + \frac{4\Delta}{3} \right] \quad (0 < -\Delta < t_0),$$

$$= \frac{\exp(-\Delta)}{3t_0} \left[-\frac{4\Delta}{3} - \frac{7t_0}{3} \right] \quad (t_0 < -\Delta < 2t_0), \qquad (5.56)$$

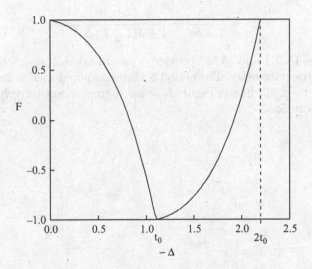

Figure 5.14: The phase-shift function F as a function of $-\Delta > 0$.

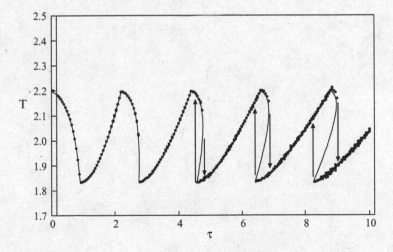

Figure 5.15: Period as a function of τ for the reduced DDE problem ($\varepsilon = 0$). The dots have been obtained numerically ($\delta = 0.1$). The line is the analytical approximation valid in the limit δ small.

where $t_0 = \ln(3)$. The function is represented in Figure 5.14. Injecting (5.54) into Eq. (5.52), we find that σ satisfies the following equation

$$\sigma = 1 - F(\Delta), \tag{5.57}$$

where Δ is defined by (5.55). This equation admits an analytical solution in parametric form. From (5.55), we obtain

$$\tau = \frac{-\Delta}{1 + \delta\sigma} = \frac{-\Delta}{1 + \delta(1 - F(\Delta))}. \tag{5.58}$$

Equations (5.57) and (5.58) provide a parametric solution for $\sigma = \sigma(\tau)$ where Δ is the parameter. The period is then computed as $T = 2\pi/(1+\delta\sigma)$. It is shown by a full line in Figure 5.15 and agrees quantitatively with the numerical solution.

6
Mechanical vibrations

6.1   Control engineering

Control engineering saw rapid development in many countries in the period immediately following the Second World War. Research groups were set up in industrial, academic, and government laboratories. Heretofore secret wartime work was widely disseminated, and new military, industrial, and other applications of the emerging discipline were identified. Alexandr Aleksandrovich Andronov (1901–1952) was a key figure in the development of control engineering in the former Soviet Union during this period, yet his name, and his contributions to control theory and nonlinear dynamics, are much less well known in the West than they deserve [30]. Major figures in the promulgation of his work in the English-speaking world include Princeton mathematician Solomon Lefschetz (e.g., in [4], [131]) and Nicholas Minorsky [164]. The latter performed an invaluable service by reporting the Soviet state of the art in a set of (at first classified) U.S. government reports, which formed the basis of later published texts. Lefschetz was convinced that the Soviet nonlinear oscillation research was a vital field of applied mathematics that had been neglected in the United States. In the context of heavy military support for basic research created by World War II and the Cold War, he set up a "project on differential equations" most specifically devoted to nonlinear equations [8], p.291.

DDE problems naturally appear in control engineering. Any system involving a feedback control will almost certainly involve time delays. These arise because a finite time is required to sense information and then react

T. Erneux, *Applied Delay Differential Equations*, Surveys and Tutorials
in the Applied Mathematical Sciences 3, DOI 10.1007/978-0-387-74372-1_6,
© Springer Science+Business Media, LLC 2009

to it. In the 1930s and 1940s, studies began of certain stabilization problems in which delay plays a role (Minorsky[1] [164]). Consider, for example, a system whose motion is described by the following second order linear equation

$$y'' + ay' + y = 0. \tag{6.1}$$

The solution of this equation with arbitrarily specified initial conditions is a function that decays exponentially toward zero. Let us assume that the solution is underdamped ($a^2 < 4$) and we wish to somehow increase the damping coefficient a in order to diminish the oscillations more rapidly. If our system is a spring-mass system then we might simply immerse the whole system in motor oil. However, if, as in Minorsky's case, our system is a ship rolling in the sea and y is the angle of tilt from the normal upright position [166] (Figure 6.1), we must be more ingenious. We might, for example, introduce ballast tanks, partially filled with water, on each side of the ship. We would also have a servomechanism designed to pump water from one tank to the other in an attempt to counteract the roll of the ship. It is hoped that this will introduce another term proportional to y' in the equation say, by' :

$$y'' + ay' + by' + y = 0. \tag{6.2}$$

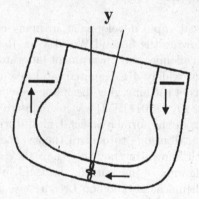

Figure 6.1: Ship rolling and its control. Two ballast tanks are partially filled with water on each side of the ship. A servomechanism pumps water from one tank to the other trying to counteract the roll of the ship (redrawn from Driver [164]).

---

[1]Nicholas Minorsky (1885–1970) contributed to several problems in ship navigation. In 1922 he presented a detailed analysis of a position feedback control system which today is refered to as Proportional-Integral-Derivative (PID) control [165]. Minorsky was born in Russia and studied at the Imperial Technical School in St. Petersburg. He joined the Russian Navy in 1917 but immigrated to the U.S. in 1918. His experiments with the U.S. Navy on PID control started in 1922. The work was successful but he didn't pursue it.

But now we need to recognize that the servomechanism cannot respond instantaneously. The control takes $\tau > 0$ to respond and therefore the control term is proportional to the velocity at the earlier instant $t - \tau$. Thus instead of Eq. (6.2), we should consider

$$y'' + ay' + by'(t - \tau) + y = 0. \tag{6.3}$$

But what would happen if the delayed control was too slow? Could the force $by'(t - \tau)$ initiate some undesired instability? We analyze this problem in the next section by assuming small values of a and b.

6.2   The method of multiple scales

Equation (6.1) with $y(0) = 0$ and $y'(0) = 1$ has an exact solution. Its expression simplifies for small a as

$$y = \exp(-at/2)\sin(t) + O(a). \tag{6.4}$$

The expression (6.4) displays the product of two functions with distinct time-scales, namely a *fast time* t and a *slow time* $s \equiv at$. This suggests seeking a solution of the DDE (6.3) for small a and $b = O(a)$ that is a function of both t and s. Specifically, we introduce a small parameter ε defined by

$$\varepsilon \equiv a \tag{6.5}$$

and expand b as

$$b = \varepsilon b_1 + ... \tag{6.6}$$

We then seek a solution of Eq. (6.3) of the form

$$y = y_0(t, s) + \varepsilon y_1(t, s) + ..., \tag{6.7}$$

where $s \equiv \varepsilon t$ is defined as a slow time variable. In the method of multiple scales [22, 124], the two times t and s are treated as independent variables. This implies the chain rules

$$y' = y_t + \varepsilon y_s$$
$$y'' = y_{tt} + 2\varepsilon y_{ts} + \varepsilon^2 y_{ss}, \tag{6.8}$$

where the subscripts mean partial derivatives with respect to t or s. The two independent time variables also mean that $y(t - \tau)$ is now a function of both $t - \tau$ and $s - \varepsilon\tau$. Expanding for small ε, we have

$$y(t - \tau) = y(t - \tau, s - \varepsilon\tau) = y(t - \tau, s) - \varepsilon\tau y_s(t - \tau, s) + ... \tag{6.9}$$

Substituting (6.5)–(6.9) into Eq. (6.3), and equating to zero the coefficients of each power of ε lead to a sequence of simple problems for the unknown functions $y_0, y_1, \ldots$. The first two problems are given by

$$O(1): y_{0tt} + y_0 = 0, \qquad (6.10)$$

$$O(\varepsilon): y_{1tt} + y_1 = -2y_{0ts} - y_{0t} - b_1 y_{0t}(t - \tau). \qquad (6.11)$$

Equation (6.10) has the solution

$$y = A(s)\exp(it) + c.c., \qquad (6.12)$$

where c.c. means complex conjugate and $A(s)$ is an unknown (complex) function of s. Substituting (6.12) into the right-hand side of Eq. (6.11), we find

$$y_{1tt} + y_1 = -i\left[2A' + A + b_1 A\exp(-i\tau)\right]\exp(it) + c.c. \qquad (6.13)$$

Because the right hand side exhibits the functions $\exp(\pm it)$ which are identical to the solutions of the homogeneous problem, y_1 will exhibit secular terms of the form $t\exp(\pm it)$. These terms become large as t increases and prevent the perturbation expansion (6.7) from being valid. Solvability of Eq. (6.13) then means that we equate to zero the coefficients of $\exp(\pm it)$ in the right hand side so that y_1 is a bounded function of t. This condition provides an ordinary differential equation for A given by

$$2\frac{dA}{ds} = -A - b_1 A\exp(-i\tau). \qquad (6.14)$$

Introducing $A = R\exp(i\phi)$ and separating real and imaginary parts, Eq. (6.14) is equivalent to the following equations for R and ϕ,

$$2\frac{dR}{ds} = -(1 + b_1\cos(\tau))R, \qquad (6.15)$$

$$2\frac{d\phi}{ds} = b_1\sin(\tau). \qquad (6.16)$$

The solution of Eqs. (6.15) and (6.16) is

$$R = R(0)\exp\left[-\frac{1}{2}(1 + b_1\cos(\tau))\right]s \qquad (6.17)$$

$$\phi = \frac{1}{2}b_1\sin(\tau)s + \phi(0) \qquad (6.18)$$

The expression (6.18) means that there is a frequency correction if $\sin(\tau) \neq 0$. Using (6.12) with $A = R\exp(i\phi)$, the frequency of the rapid oscillations in t is now given by

$$\omega(\varepsilon) = 1 + \frac{\varepsilon}{2}b_1\sin(\tau). \qquad (6.19)$$

On the other hand, the expression (6.17) means that R slowly decays exponentially if

$$\Gamma \equiv 1 + b_1 \cos(\tau) > 0. \tag{6.20}$$

If τ progressively increases from zero, the damping rate Γ decreases from $1 + b_1$ and becomes less than 1 if $\cos(\tau) < 0$. Moreover, if both $\cos(\tau) < 0$ and $b_1 > 1$, Γ may even change sign meaning that R will grow exponentially in time. Minorsky was fully aware of this problem and investigated the response of the nonlinear mechanical oscillator in detail.

6.3 Minorsky's equation

Minorsky assumed that the delayed feedback is nonlinear and leads to a nonlinear damping term of the form $-by' + \varepsilon c y'^3$. Instead of the linear DDE (6.3), he then analyzed the following equation,

$$y'' + ay' + y = -by'(t - \tau) + \varepsilon c y'^3(t - \tau), \tag{6.21}$$

which is called Minorsky's equation [192]. The coefficients $\varepsilon = a \ll 1$, b, and c are positive. Pinney [192] discussed another problem (sound generated by a speaker) described by the same equation.

6.3.1 Hopf bifurcation

The analysis of Eq.(6.21) is similar to our previous analysis of Eq. (6.3) and we summarize the main results. The leading solution of Eq.(6.21) is again given by (6.12). But the solvability condition of the $O(\varepsilon)$ problem leads to a new amplitude equation given by

$$2\frac{dA}{ds} = -A - b_1 A \exp(-i\tau) + 3cA^2 \overline{A} \exp(-i\tau). \tag{6.22}$$

Introducing $A = R \exp(i\phi)$ into Eq. (6.22), we obtain the following equations for R and ϕ,

$$2\frac{dR}{ds} = -(1 + b_1 \cos(\tau))R + 3cR^3 \cos(\tau) \tag{6.23}$$

$$2\frac{d\phi}{ds} = b_1 \sin(\tau) - 3cR^2 \sin(\tau). \tag{6.24}$$

Equation (6.23) is an equation for R only. Knowing R, we determine the phase ϕ by integrating Eq. (6.24). From Eq. (6.23), we note that if

$$\Gamma \equiv 1 + b_1 \cos(\tau) < 0, \tag{6.25}$$

the steady-state solution $R = 0$ is unstable. Mathematically, the critical point $b_1 = b_{1H}$ where

$$b_{1H} \equiv -\cos^{-1}(\tau) \ (\cos(\tau) < 0) \tag{6.26}$$

is a Hopf bifurcation point because $y(t)$ will exhibit oscillations as soon as $b_1 > b_{1H}$. The delay τ needs to be large enough in order to verify the condition $\cos(\tau) < 0$. The question now is whether the oscillations will continuously grow in time or if they will reach a fixed amplitude $R \neq 0$. Setting $dR/ds = 0$ in Eq. (6.23) and solving for R, we find

$$R = \sqrt{\frac{b_1 - b_{1H}}{3c}} \tag{6.27}$$

if $b_1 \geq b_{1H}$. This solution is the Hopf bifurcation branch that emerges from the Hopf bifurcation point $b_1 = b_{1H}$. Its amplitude has the typical square-root behavior of a Hopf bifurcation. Because $y \simeq R \exp(i(t + \phi)) + c.c.$, the extrema of the oscillations are given by

$$y \simeq \pm 2R. \tag{6.28}$$

Finally, we may investigate the stability of this solution. Linearizing Eq. (6.23) for $R \neq 0$ and simplifying, the small perturbation u satisfies

$$\frac{du}{ds} = 3cR^2 \cos(\tau)u. \tag{6.29}$$

Because $\cos(\tau) < 0$ and $c > 0$, the growth rate $3cR^2 \cos(\tau)$ is negative and the Hopf bifurcation branch is stable. The Hopf theorem [100] stating that a supercritical bifurcation (here, $b_1 > b_{1H}$) is stable is verified.

Figure 6.2 shows the numerical bifurcation diagram of Minorsky's equation (6.21) for the following values of the parameters

$$a = 0.1, \quad c = 1, \quad \varepsilon = 0.1, \quad \text{and} \quad \tau = 3\pi. \tag{6.30}$$

From (6.26), we determine $b_{1H} = 1$. Using then (6.27) and (6.28), we determine the extrema of y as $y = \pm 2\sqrt{(b_1 - 1)/3}$. The two parabolic lines are shown in Figure 6.2. They are in good agreement with the numerical branches until a secondary bifurcation appears at $b_S = 0.35$. This new bifurcation is a bifurcation to quasiperiodic oscillations (torus bifurcation). These oscillations are characterized by two noncommensurable frequencies, namely $\omega_1 \simeq 1$ and $\omega_2 = O(\varepsilon)$.

6.3.2 Hopf bifurcation for large delay

We suspect that the torus bifurcation results from the relatively large value of τ (here $\tau = 3\pi \simeq 9$). But if we now consider τ as a large parameter, we need to compare it to ε^{-1}. We also need to be more careful as we expand the delayed variable $y(t - \tau)$. Recall that Eq. (6.22) is derived assuming $\tau = O(1)$. The slow time delay

$$\theta \equiv \varepsilon\tau \tag{6.31}$$

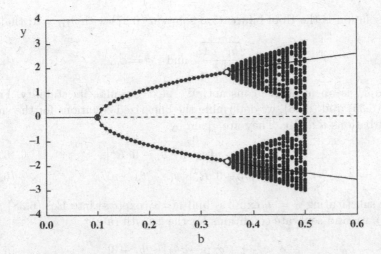

Figure 6.2: Bifurcation diagram of Minorsky's equation. A Hopf bifurcation appears at $b_H \equiv \varepsilon b_{1H} = 0.1$ and is followed by a secondary bifurcation to quasiperiodic oscillations at $b_S \simeq 0.35$. Other bifurcations appear for higher values of b. The two parabolic lines are the approximations of the Hopf bifurcation branch.

is a small $O(\varepsilon)$ quantity and we may expand $y(t - \tau, s - \varepsilon\tau)$ in power series of ε as in (6.9). However, this expansion is no longer valid if $\tau = O(\varepsilon^{-1})$ or larger. Nevertheless, the leading expression of the solution is still given by (6.12) and the solvability condition now leads to

$$2\frac{dA}{ds} = -A - A(s - \theta)\exp(-i\tau)\left[b_1 - 3c\,|A(s - \theta)|^2\right], \qquad (6.32)$$

where we note the appearance of the slow time delayed amplitude $A(s - \theta)$. In contrast to Eq. (6.22), Eq. (6.32) is now a DDE. This DDE is, however, an important simplification of the original DDE problem because steady and periodic solutions of Eq. (6.32) correspond to periodic and quasiperiodic solutions of Eq. (6.21), respectively. For mathematical simplicity, we analyze Eq.(6.32) with

$$\tau = (1 + 2n)\pi, \qquad (6.33)$$

where n is an arbitrary large integer. Equation (6.32) then simplifies as

$$2\frac{dA}{ds} = -A + A(s - \theta)\left[b_1 - 3c\,|A(s - \theta)|^2\right]. \qquad (6.34)$$

Introducing $A = R\exp(i\phi)$ into Eq. (6.34), we obtain

$$2R' = -R + R(s - \theta)\left[b_1 - 3cR^2(s - \theta)\right]\cos(\phi(s - \theta) - \phi), \qquad (6.35)$$

$$2R\phi' = R(s - \theta)\left[b_1 - 3cR^2(s - \theta)\right]\sin(\phi(s - \theta) - \phi). \qquad (6.36)$$

One solution is the Hopf bifurcation solution (6.27) with $b_{1H} = 1$; that is,

$$R = \sqrt{\frac{b_1 - 1}{3c}} \quad \text{and} \quad \phi = \phi_0, \tag{6.37}$$

where ϕ_0 is an arbitrary constant. We next examine its stability. From Eqs. (6.35) and (6.36), we determine the linearized equations for the small perturbations u and v. They are given by

$$2u' = -u + u(s - \theta)\left[b_1 - 9cR^2\right], \tag{6.38}$$
$$2v' = \left[b_1 - 3cR^2\right](v(s - \theta) - v). \tag{6.39}$$

After substituting $u = u_0 \exp(\sigma s)$ and $v = v_0 \exp(\sigma s)$ into Eqs.(6.38) and (6.39), we find separate conditions for the growth rate σ

$$2\sigma = -1 + \exp(-\sigma\theta)\left[-2b_1 + 3\right], \tag{6.40}$$
$$2\sigma = \exp(-\sigma\theta) - 1. \tag{6.41}$$

A graphical analysis of (6.41) (we study the intersections of the functions 2σ and $\exp(-\sigma\theta) - 1$) indicates that $\sigma = 0$ is the only solution. Assuming $\sigma \neq 0$, we concentrate on Eq. (6.40) and wonder if a Hopf bifurcation is possible. Inserting $\sigma = i\omega$ into Eq. (6.40), we find two conditions from the real and imaginary parts. They are given by

$$0 = -1 + \cos(\omega\theta)(-2b_1 + 3), \tag{6.42}$$
$$2\omega = -\sin(\omega\theta)(-2b_1 + 3). \tag{6.43}$$

We may eliminate b_1 and obtain a single equation for ω. With Eq. (6.42), we then obtain the conditions

$$\tan(\omega\theta) = -2\omega, \tag{6.44}$$
$$2b_1 = 3 - 1/\cos(\omega\theta). \tag{6.45}$$

Using the values of the parameters listed in (6.30), we determine $\theta = \varepsilon\tau \simeq 0.94$. Solving then Eqs. (6.44) and (6.45) numerically, we obtain

$$\omega\theta \simeq 1.8 \quad \text{and} \quad b_1 \simeq 3.5. \tag{6.46}$$

The point $(b, \pm 2R)$ where $b = \varepsilon b_1 = 0.35$ and $2R = 2\sqrt{(b_1 - 1)/3} = 1.825$ is shown in Figure 6.2 by two open circles. They exactly match the numerical estimates of the torus bifurcation point obtained by integrating Eq. (6.21).

6.3.3 Torus bifurcation

At the torus bifurcation point, $u = u_0 \exp(iws) + c.c.$ but $v = 0$. This suggests we seek a particular solution with $R = R(s) \neq 0$ but $\phi = \phi_0$

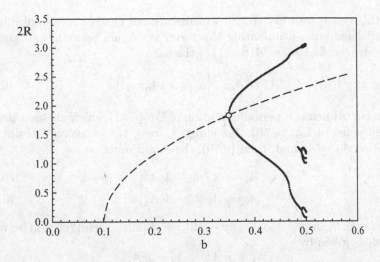

Figure 6.3: Bifurcation diagram of the slow time amplitude equation. The broken line is the Hopf bifurcation branch $2R = \sqrt{(b-\varepsilon)/3\varepsilon}$. The full line is the extremum of $2R$ obtained numerically.

constant. From Eqs. (6.35) and (6.36), we then find that R satisfies a scalar DDE given by

$$2R' = -R + R(s-\theta)\left[b_1 - 3cR^2(s-\theta)\right]. \tag{6.47}$$

The bifurcation diagram of Eq. (6.47) is shown in Figure 6.3. The figure indicates a Hopf bifurcation of Eq. (6.47) corresponding to a torus bifurcation of the original equation (6.21). Slightly before $b_1 = 0.5$, a new bifurcation appears which we do not analyze.

6.3.4   Map

We wish to determine an analytical approximation of the torus bifurcation branch for large θ. A graphical analysis of Eq. (6.44) (look for the intersections of $\tan(\omega\theta)$ and the straight line $-(\omega\theta)/\theta$) indicates that the first positive root of Eq. (6.44) ($\pi/2 < \omega\theta < \pi$) approaches $\omega\theta = \pi$ as $\theta \to \infty$. Equation (6.45) then gives b_1 as

$$\omega\theta = \pi \quad \text{and} \quad b_1 = 2. \tag{6.48}$$

Equations (6.48) suggest that the new frequency ω scales as θ^{-1}. We take this feature into account by introducing the following new slow time variable

$$S \equiv s/\theta. \tag{6.49}$$

Inserting (6.49) into Eq. (6.32), we note a small $O(\theta^{-1})$ term multiplying the left hand side. Eliminating this term, we obtain an equation relating $A_n = A(S)$ and $A_{n-1} = A(S-1)$ of the form

$$A_n = A_{n-1}\left[b_1 - 3c\,|A_{n-1}|^2\right]. \qquad (6.50)$$

The torus branch is a periodic solution of Eq. (6.32) which means a period 2 fixed point of Eq. (6.50). Assuming A_n real, this solution satisfies the condition $A_2 = A_0$ and, from (6.50), the conditions

$$A_1 = A_0\left[b_1 - 3cA_0^2\right], \qquad (6.51)$$
$$A_0 = A_1\left[b_1 - 3cA_1^2\right]. \qquad (6.52)$$

Eliminating b_1 in Eqs. (6.51) and (6.52), we obtain a useful relation between A_1 and A_0 given by

$$\frac{A_1 + 3cA_0^3}{A_0} = \frac{A_0 + 3cA_1^3}{A_1}. \qquad (6.53)$$

This equation can be rewritten as

$$(A_1^2 - A_0^2)(1 - 3cA_1A_0) = 0. \qquad (6.54)$$

The first solution of Eq. (6.54) is $A_1^2 = A_0^2$ and represents the steady state. The second solution of Eq. (6.54) satisfies $1 - 3cA_1A_0 = 0$ which implies that

$$A_1 = \frac{1}{3cA_0}. \qquad (6.55)$$

From (6.51), we then obtain $1/(3cA_0) = A_0\left[b_1 - 3cA_0^2\right]$ which leads to the quadratic equation

$$9c^2A_0^4 - 3cb_1A_0^2 + 1 = 0. \qquad (6.56)$$

This equation admits a real solution only if $b_1 \geq 2$. In implicit form, it is given by

$$b_1 - 2 = \frac{3c}{A_0^2}(A_0^2 - A_{0s}^2)^2, \qquad (6.57)$$

where $A_{0s}^2 \equiv 1/(3c)$. The secondary bifurcation branch is clearly super-critical because $b_1 > 2$ if $A_0^2 \neq A_{0s}^2$. The first and secondary bifurcation branches are shown in Figure 6.4 and compared to the solutions obtained numerically from Eq. (6.50).

A second example of a nonlinear second-order DDE is examined in the next section.

6.3.5 Large delay asymptotics: Ginzburgh–Landau equation

Our previously analysis of Minorsky's equation (6.21) is valid provided $\varepsilon \to 0$ and $\tau = O(\varepsilon^{-1})$. The solution of the amplitude DDE quantitatively

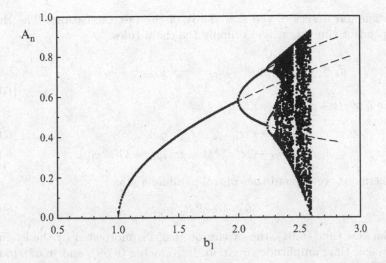

Figure 6.4: Bifurcation diagram of the slow time amplitude equations for large delay. The broken lines are the primary and secondary branches obtained analytically. The figure indicates a period-4 bifurcation followed by higher-order bifurcations that we do not analyze.

agrees with the numerical solution of the original equation suggesting that the amplitude DDE could be valid for larger values of τ. In this section, we show that this conclusion is premature because a distinguishing limit is possible if $\tau = O(\varepsilon^{-2})$.

Specifically, we consider discrete values of τ of the form (6.33) with n large and scale τ as

$$\tau = \varepsilon^{-2}\tau_0. \tag{6.58}$$

Restricting τ to (6.33) is not a limitation of the multiple-scale method but allows that the first Hopf bifurcation be simply located at $b = a = \varepsilon$. We then seek a small amplitude solution of the form

$$y = \varepsilon y_1(t,x,\nu) + \varepsilon^2 y_2(t,x,\nu) + \dots \tag{6.59}$$

where

$$x \equiv \tau^{-1}(1 + \varepsilon c_1 + \varepsilon^2 c_2 + \dots)t, \tag{6.60}$$

$$\nu \equiv \varepsilon^2 \tau^{-1} t, \tag{6.61}$$

are two slow-time variables. The strained coordinate x is motivated by the large delay τ. The corrections $c_1, c_2, \dots$ will be determined by solvability conditions. The strained coordinate eliminates the need for an additional slow time $\varepsilon \tau^{-1} t$. The variable ν is motivated by the leading-order amplitude equation and describes the slow decay of the fast oscillations. The three independent times t, x, and ν imply the chain rules

$$y_t(t - \tau) = \bar{y}_t - (\varepsilon c_1 + \varepsilon^2 c_2 + ...)\bar{y}_{tx} + \frac{\varepsilon^2 c_1^2}{2}\bar{y}_{txx} - \varepsilon^2 \bar{y}_{t\nu} + \varepsilon^2 \tau_0^{-1}\bar{y}_x + O(\varepsilon^3),$$
(6.62)

where $\bar{y} \equiv y(t - \tau, x - 1, \nu)$ and

$$y' = y_t + \varepsilon^2 \tau_0^{-1} y_x + O(\varepsilon^3 y_x),$$
(6.63)

$$y'' = y_{tt} + 2\varepsilon^2 \tau_0^{-1}(1 + \varepsilon c_1)y_{tx} + O(\varepsilon^4 y_{tx}).$$
(6.64)

Furthermore, we expand the control parameter b as

$$b = \varepsilon + \varepsilon^2 \beta_2 + ...$$
(6.65)

As the slow time (6.61), the expansion (6.65) is motivated by the leading-order slow time amplitude equation. Introducing (6.59), and (6.62)–(6.65) into Eq. (6.21) with $a = \varepsilon$, we obtain the following sequence of problems for the unknown functions $y_1, y_2, ...,$

$$O(\varepsilon) : y_{1tt} + y_1 = 0,$$
(6.66)

$$O(\varepsilon^2) : y_{2tt} + y_2 = -y_{1t} - \bar{y}_{1t},$$
(6.67)

$$O(\varepsilon^3) : y_{3tt} + y_3 = -y_{2t} - \bar{y}_{2t} + c_1\bar{y}_{1tx} - 2\tau_0^{-1}y_{1tx},$$
(6.68)

$$O(\varepsilon^4) : y_{4tt} + y_4 = -y_{3t} - \bar{y}_{3t} + c_1\bar{y}_{2tx} - 2\tau_0^{-1}y_{2tx}$$
$$+ c_2\bar{y}_{1tx} - \left[\frac{c_1^2}{2}\bar{y}_{1txx} - \bar{y}_{1t\nu} + \tau_0^{-1}\bar{y}_{1x}\right]$$
$$- \tau_0^{-1}y_{1x} - \beta_2\bar{y}_{1t} - 2c_1\tau_0^{-1}y_{1tx} + c\bar{y}_{1t}^3.$$
(6.69)

The solution of Eq. (6.66) is

$$y_1 = A(x, \nu)\exp(it) + c.c.,$$
(6.70)

where A is an unknown complex amplitude. Solvability of Eq. (6.67) with respect to the periodic solution $\exp(\pm it)$ requires the condition

$$-iA - iA(x - 1)\exp(-i\tau) = -i[A - A(x - 1)] = 0,$$
(6.71)

or equivalently,

$$A(x - 1) = A.$$
(6.72)

The solution of Eq. (6.67) is then

$$y_2 = B(x, \nu)\exp(it) + c.c.,$$
(6.73)

where B is a new unknown amplitude. The solvability condition for Eq. (6.68) now leads to the following equation for B,

$$i[B - B(x - 1)] = c_1 iA_x(x - 1)\exp(-i\tau) - 2i\tau_0^{-1}A_x.$$
(6.74)

This equation defines an equation for a map of the form for $B_m - B_{m-1} = F$, where F is a constant. A bounded solution for B requires that $F = 0$ or

$$c_1 i A_x(x-1) \exp(-i\tau) - 2i\tau_0^{-1} A_x = -i A_x \left(c_1 + 2\tau_0^{-1}\right) = 0. \qquad (6.75)$$

Equation (6.75) is satisfied if

$$c_1 = -2\tau_0^{-1}. \qquad (6.76)$$

The solution of Eq. (6.68) is then

$$y_3 = C(x, \nu) \exp(it) + c.c. \qquad (6.77)$$

and B satisfies the condition $B(x-1) = B$. Because A is still unknown, we explore Eq. (6.69). Applying the two previously discussed solvability conditions now leads to the following equation (after simplifying using (6.58) and (6.72))

$$0 = -c_2 i A_x - \left[-\frac{c_1^2}{2} i A_{xx} + i A_\nu - \tau_0^{-1} A_x\right]$$
$$- \tau_0^{-1} A_x + \beta_2 i A - 2c_1 \tau_0^{-1} i A_x - 3i c A^2 \overline{A}. \qquad (6.78)$$

We eliminate the term multiplying A_x by requiring that c_2 satisfies $c_2 + 2c_1\tau_0^{-1} = 0$. The remaining terms then lead to the following Ginzburg–Landau equation [48]

$$A_\nu = 2\tau_0^{-2} A_{xx} + \beta_2 A - 3A^2 \overline{A}. \qquad (6.79)$$

with A satisfying the boundary condition (6.72). The singular nature of the delay now clearly appears in the diffusion coefficient that decreases like τ_0^{-2} as $\tau_0 \to \infty$. τ_0 plays the same role as the length of a spatiotemporal system. If τ_0 is sufficiently large, stable periodic solutions in x are possible [229].

A historical note is in order. Physicists have long been fascinated by the idea that a DDE could be equivalent to a partial differential equation (PDE). Giacomelli et al.[72] and Grigorieva et al. [80] proposed that a small amplitude periodic solution of a DDE could be captured by a Ginzburg–Landau equation [48] that exhibits either identical perturbation solutions or identical spectral properties. The question of deriving such an equation from solvability conditions directly applied to the original DDE remained open. The subsequent works by Nizette [178], Erneux et al. [61], and Wolfrum and Yanchuk [244] on first-order DDEs then showed that two successive solvability conditions can be used to properly derive the expected PDE.

6.4 Johnson and Moon's equation

Johnson and Moon [115] investigated experimentally an electromechanical system. Their experiments were compared to the numerical solutions of the following equation

$$y'' + ay' + b(y - y^3) = c(y' - y'(t-1)) \tag{6.80}$$

which exhibits periodic, quasiperiodic, and chaotic oscillations. The values of the parameters were

$$a = 2.623 \quad \text{and} \quad b = 170\pi^2 \tag{6.81}$$

with c the control parameter. The bifurcation diagram of the stable solutions is shown in Figure 6.5 and reveals both Hopf and torus bifurcations. We take into account the fact that b is large by introducing the new time $T \equiv b^{1/2}t$. Equation (6.80) then becomes

$$y'' + y - y^3 = \varepsilon^2(-ay' + c(y' - y'(T - \varepsilon^{-2}))), \tag{6.82}$$

where prime now means differentiation with respect to T and

$$\varepsilon^2 \equiv b^{-1/2} \tag{6.83}$$

is a small parameter.

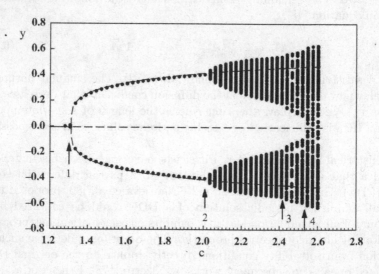

Figure 6.5: Bifurcation diagram. The values of the fixed parameters are $a = 2.623$ and $b = (13\pi)^2$. The broken lines correspond to the approximation of the Hopf bifurcation branch. Labels 1 to 4 mark succesive bifurcations.

6.4.1 Unusual Hopf bifurcation

The left-hand side is the equation of a conservative oscillator whereas the right-hand side is responsible for $O(\varepsilon^2)$ small damping. The small parameter suggests seeking a small amplitude solution. To this end, we use the method of multiple time-scales introduced previously in Section 6.2. Specifically, we seek a solution of the form

$$y = \varepsilon y_1(T, s) + \varepsilon^3 y_3(T, s) + ..., \tag{6.84}$$

where $s \equiv \varepsilon^2 T$. The power series is an odd power of ε because the only nonlinear term in Eq. (6.84) is the cubic function y^3. Because of the two-time formulation, we need to rewrite the delayed term $y'(T - \varepsilon^{-2})$ as

$$y_T(T - \varepsilon^{-2}) = y_T(T - \varepsilon^{-2}, s - 1) + \varepsilon^2 y_s(T - \varepsilon^{-2}, s - 1). \tag{6.85}$$

Introducing (6.84) and (6.85) into Eq. (6.82) leads to the following problems for the functions y_1 and y_3,

$$y_{1TT} + y_1 = 0, \tag{6.86}$$

$$y_{3TT} + y_3 = y_1^3 - ay_{1T} + c\left[y_{1T} - y_{1T}(T - \varepsilon^{-2}, s - 1)\right]$$
$$- 2y_{1Ts}. \tag{6.87}$$

The solution of Eq. (6.86) is

$$y_1 = A(s)\exp(iT) + c.c. \tag{6.88}$$

Solvability of Eq. (6.87) then implies the condition

$$3A^2\overline{A} - iaA + ic\left[A - A(s - 1)\exp(-i\varepsilon^{-2})\right] - 2iA' = 0. \tag{6.89}$$

In terms of $A = R\exp(i\phi)$, the real and imaginary parts of Eq. (6.89) give

$$2R' = -aR + c(R - R(s - 1)\cos(\varepsilon^{-2} - \phi(s - 1) + \phi)), \tag{6.90}$$

$$2R\phi' = -3R^3 + cR(s - 1)\sin(\varepsilon^{-2} - \phi(s - 1) + \phi)). \tag{6.91}$$

The Hopf bifurcation branch satisfies the condition $R = cst$ and $\phi = \sigma s$ where σ is the correction of the Hopf frequency. From Eqs. (6.90) and (6.91), we obtain

$$0 = -a + c(1 - \cos(\varepsilon^{-2} + \sigma)), \tag{6.92}$$

$$2\sigma = -3R^2 + c\sin(\varepsilon^{-2} + \sigma) \tag{6.93}$$

which leads to the parametric solution

$$c = \frac{a}{(1 - \cos(\varepsilon^{-2} + \sigma))}, \tag{6.94}$$

$$R^2 = \frac{1}{3}(-2\sigma + c\sin(\varepsilon^{-2} + \sigma)) \geq 0. \tag{6.95}$$

In Figure 6.5, we compare this solution to the numerical solution. The value of b is $b = (13\pi)^2$ implying $\varepsilon^{-2} = 13\pi$. The expressions (6.94) and (6.95) then simplify as

$$c = \frac{a}{(1 + \cos(\sigma))}, \tag{6.96}$$

$$R^2 = -\frac{1}{3}(2\sigma + c\sin(\sigma)) \geq 0, \tag{6.97}$$

where $-\pi/2 < \sigma \leq 0$. The Hopf bifurcation point corresponds to $\sigma = 0$ and $R = 0$ and is located at

$$c_H \equiv a/2. \tag{6.98}$$

We may also expand (6.96) and (6.97) for small σ. Then eliminating σ, we obtain R as a function of $c - c_H$ as

$$R = \sqrt{\frac{2}{3}(2 + c_H)} \left(\frac{c - c_H}{c_H}\right)^{1/4} \tag{6.99}$$

which exhibits a $(c - c_H)^{1/4}$ power law that differs from the traditional squareroot law. As for Minorsky's equation, we may now analyze the stability of the Hopf bifurcation branch and possibly determine the secondary Hopf bifurcation. In Figure (6.6), we determine numerically the bifurcation

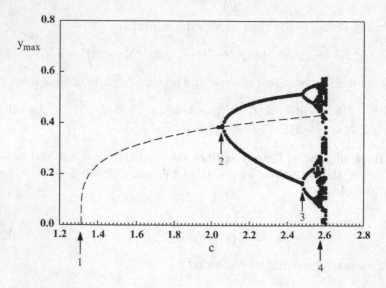

Figure 6.6: Bifurcation diagram of the slow time amplitude equations. The broken line comes from the expression of the Hopf bifurcation branch. Labels 1 to 4 mark the same bifurcations as in Figure 6.5.

diagram of the slow time amplitude equations (6.90) and (6.91). The succesive bifurcations compare well with the bifurcations determined from the orginal equation (compare Figures 6.5 and 6.6).

6.5 Machine tool vibrations

The history of machine tool chatter goes back almost 100 years to when machine tool chatter was recognized as one of the most delicate problems facing the machinist [217]. Today, the so-called regenerative effect has become the commonly accepted explanation for machine tool chatter. Chatter instability is a violent vibratory motion occurring between the cutting tool and workpiece when the tool is removing material in the form of chips from a previously machined surface profile. Its presence in machining operations has been estimated to be the most undesirable factor increasing manufacturing costs of vital components for aerospace, automotive, biomedical construction, household devices, military, transportation, and many other machine-related applications. See Figure 6.7. Following a small external perturbation, the tool starts a damped vibration relative to the workpiece. The surface of the workpiece becomes wavy and after a round, the chip thickness will vary at the tool because of this wavy surface. Thus, the cutting force depends on the actual and delayed values of the relative displacement of the tool and workpiece. The delay is exactly equal to the time of revolution of the workpiece and is the key source of the regenerative effect allowing self-excited vibrations in the machining operation. The energy for the vibrations comes from the forward motion of the tool and workpiece. The frequency is typically slightly larger than the natural frequency of the most flexible vibration mode of the machine-tool system. The frequency is

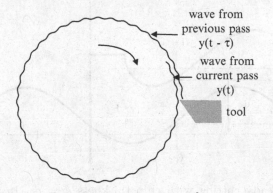

wave from
previous pass
$y(t - \tau)$

wave from
current pass
$y(t)$

tool

Figure 6.7: Workpiece (left) and cutting tool (right) are vibrating because of the wavy surface that is never the same after each round of the workpiece.

not synchronous with the spindle drive frequency. This mismatch typically accounts for the screechy sound that is characteristic of chatter.

6.5.1   Formulation and stability

In order to describe the instability, we assume orthogonal cutting and consider the problem as one-dimensional. The machine tool is characterized by its mass m, spring constant k, and damping r. The zero value of the tool edge position is set in a way that the cutting force F_c is in balance with the spring force when the chip thickness d is just the prescribed steady-state cutting value d_0 (see Figures 6.8 and 6.9). The restoring force of the

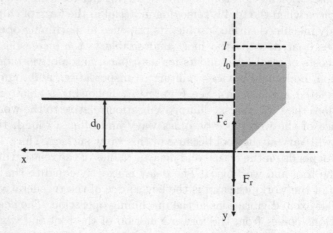

Figure 6.8: Steady-state cutting.

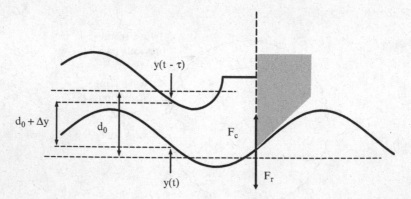

Figure 6.9: Steady cutting with chip thickness $d = d_0$. In practice, the thickness varies with time and is a function of the present and previous position of the tool: $d = d_0 + \Delta y$, where $\Delta y \equiv y(t) - y(t - \tau)$.

tool is given by $F_r(y) \equiv -k(l - l_0 + y)$ where l and l_0 represent the initial spring length and spring length in steady state cutting, respectively. The cutting force $F_c(d)$ is assumed to be a function of the chip thickness only. In steady-state cutting, $F_r(0) + F_c(d_0) = 0$ which then leads to the relation

$$k(l - l_0) + F_c(d_0) = 0. \tag{6.100}$$

Equation (6.100) provides an expression for $l - l_0 < 0$ (pre-stress in the spring). The equation of motion of the tool is

$$my'' = -k(l - l_0 + y) - ry' - F_c(d) \tag{6.101}$$

which simplifies, using (6.100), as

$$my'' = -ky - ry - (F_c(d) - F_c(d_0)). \tag{6.102}$$

Dividing by m, Eq. (6.102) can be rewritten as

$$y'' + 2\zeta\omega y' + \omega_0^2 y = -\frac{1}{m} [F_c(d(t)) - F_c(d_0)], \tag{6.103}$$

where $\omega_0 \equiv \sqrt{k/m}$ is the natural angular frequency of the undamped free oscillating tool and $\zeta \equiv r/(2m\omega_0)$ is the so-called relative damping factor. The cutting force $F_c(d)$ is a strong nonlinear function of d (see Fig. 6.10) and a power-law relation of the form

$$F_c(d) = \begin{matrix} 0 & \text{if } d \leq 0 \\ Kwd^\alpha & \text{if } d > 0 \end{matrix} \tag{6.104}$$

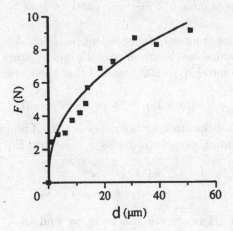

Figure 6.10: Experimental force in the feed direction for aluminum (from Kalmár-Nagy Ph.D. thesis with permission [121]).

is often used. In (6.104), w is the chip width and K is a material-based constant. For aluminum, the exponent is $\alpha = 0.41$. If $d \leq 0$, the tool is no longer in contact with the workpiece and $F_c = 0$. If the deviation $d - d_0$ is small, we may use a linear approximation and express the change in force as

$$F_c(d) - F_c(d_0) = F'_c(d_0)(d - d_0). \tag{6.105}$$

We now need to relate d and y. See Figure 6.9. The actual chip thickness depends on the present and a previous displacement of the tool at time $t - \tau$ and is given by

$$d = d_0 + y(t) - y(t - \tau), \tag{6.106}$$

where $\tau \equiv 2\pi/\Omega$ and Ω is the angular frequency of the rotating workpiece. With (6.105) and (6.106), Eq. (6.103) becomes

$$y'' + 2\zeta\omega_0 y' + \omega_0^2 y + \frac{F'_c(d_0)}{m}(y - y(t - \tau)) = 0. \tag{6.107}$$

We now formulate a dimensionless equation of motion by introducing the new time

$$s = \omega_0 t. \tag{6.108}$$

Equation (6.107) then simplifies as

$$y'' + ay' + y + b(y - y(s - \theta)) = 0, \tag{6.109}$$

where prime now means differentiation with respect to time s and the new parameters a, b, and θ are defined by

$$a \equiv 2\zeta, \quad b \equiv \frac{F'_x(d_0)}{m\omega_0^2}, \quad \text{and} \quad \theta \equiv \omega_0 \tau. \tag{6.110}$$

We have reduced the number of independent parameters from four to three.

We next wish to analyze the stability of the steady-state $y = 0$. Introducing $y = y_0 \exp(\sigma t)$ into Eq. (6.109), we find that the growth rate σ satisfies

$$\sigma^2 + a\sigma + 1 + b(1 - \exp(-\sigma\theta)) = 0. \tag{6.111}$$

We next assume that the stability boundary is a Hopf bifurcation boundary where σ is purely imaginary. Introducing $\sigma = i\omega$ into Eq. (6.111), we find

$$1 - \omega^2 + b(1 - \cos(\omega\theta)) = 0, \tag{6.112}$$
$$a\omega + b\sin(\omega\theta) = 0. \tag{6.113}$$

By eliminating the trigonometric functions, we find

$$b = \frac{(1 - \omega^2)^2 + a^2\omega^2}{2(\omega^2 - 1)}. \tag{6.114}$$

Because $b > 0$, Eq. (6.114) implies $\omega > 1$. By using the trigonometric identity

$$\frac{1 - \cos(\omega\theta)}{\sin(\omega\theta)} = \tan(\frac{\omega\theta}{2}), \qquad (6.115)$$

we may eliminate b in Eqs. (6.112) and (6.113). The resulting equation leads to an equation for θ given by

$$\theta = \frac{2}{\omega}\left[\arctan(\frac{1 - \omega^2}{a\omega}) + n\pi\right], \qquad (6.116)$$

where $n = 1, 2, \ldots$ is the wave-number of the workpiece surface modulation ($n > 0$ because $\omega > 1$ and τ is positive). The expressions (6.113) and (6.116) provide a parametric solution for $b = b(\theta)$. The dimensionless delay θ is inversely proportional to the angular velocity $\Omega_\theta \equiv 2\pi/\theta$. See Fig. 6.11. If we are chattering, we may speed the spindle up (increase Ω_θ) or change the dynamics of the tool (decrease b). The minima of the stability boundary satisfy the condition $db/d\omega = 0$. This leads to $\omega_{\min} = \sqrt{1 + a}$ and the minimum

$$b_{\min} = a(1 + a/2). \qquad (6.117)$$

Note that the analysis of Eqs. (6.112) and (6.113) for a small is more immediate. Equation (6.113) suggests the scaling $b = ab_1$ and Eq. (6.112) then motivates the scaling $\omega - 1 = a\omega_1$. Inserting these expresions of b and ω into Eqs. (6.112) and (6.113) leads to the conditions

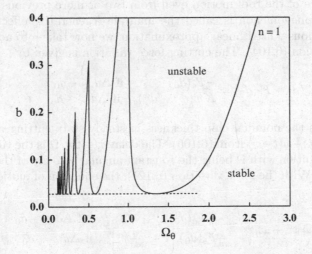

Figure 6.11: Stability lobes. The succesive maxima occurs close to $\Omega_\theta = 1/n$. They are points of resonance between the tool and workpiece oscillations implying maximal stability. $a = 0.0272$.

$$-2\omega_1 + b_1(1 - \cos(\theta)) = 0, \tag{6.118}$$

$$1 + b_1 \sin(\theta) = 0 \tag{6.119}$$

in first approximation ($a \to 0$). Equation (6.119) gives the Hopf stability boundaries; that is, $b_1 = -1/\sin(\theta) > 0$. The minima of each stability lobe occur at $b = a$ and

$$\theta = \theta_n \equiv (2n + 1)\pi - \pi/2, \tag{6.120}$$

where $n = 1, 2, \dots$. The Hopf bifurcation for the case $n = 7$ is examined in Section 6.6.2.

6.6   Experiments

Shi and Tobias [214] observed experimentally that finite-amplitude oscillations of the tool may appear as an alternate to stable stationary cutting. This suggests the coexistence of stable steady and periodic motions resulting from a subcritical Hopf bifurcation. The problem has been analyzed in detail by Kalmár-Nagy [120, 121] who showed experimentally, analytically, and numerically the hysteretic nature of the instability. His model equations incorporate both the nonlinear cutting force and the multiple-regenerative effect due to the tool leaving the cut. When the chatter amplitude exceeds a certain amplitude, the contact ceases between the tool and the workpiece and the tool starts a damped vibration until it comes in contact with the workpiece again. When this happens, the uncut chip thickness is affected by the trace of the tool motion even from two or more previous turns before. This phenomenon is called the multiple regenerative effect. Instead of the previously used linear approximation, we now take into account the full expression (6.104). The cutting force variation is given by

$$F_c(d) - F_c(d_0) = \begin{array}{ll} -F_c(d_0) & \text{if } \Delta d \le -d_0 \\ Kw(d^\alpha - d_0^\alpha) & \text{if } \Delta d > -d_0 \end{array} \tag{6.121}$$

where d_0 is the nominal chip thickness in steady-state cutting and $\Delta d \equiv d - d_0 = y(t) - y(t - \tau)$ from (6.106). The delay $\tau \equiv 2\pi/\Omega$ is the time period of one revolution with Ω being the constant angular velocity of the rotating workpiece. With the force variation (6.121), the equation of motion (6.103) now is

$$y'' + 2\zeta\omega_0 x' + \omega_0^2 x = \begin{array}{ll} -\frac{1}{m}F_c(d_0) & \text{if } \Delta d \le -d_0 \\ \frac{1}{m}F_c(d_0)\left[1 - (\frac{d}{d_0})^\alpha\right] & \text{if } \Delta d > -d_0 \end{array} \tag{6.122}$$

Introducing the dimensionless time and space variables

$$s = \omega_0 t \quad \text{and} \quad z = y/y_0, \tag{6.123}$$

where $y_0 \equiv 3d_0/(2 - \alpha)$, Eqs. (6.122) can be rewritten as

$$z'' + az' + z = \begin{cases} -\frac{b(2-\alpha)}{3\alpha} & \text{if } \Delta z \leq -\frac{2-\alpha}{3} \\ \frac{b(2-\alpha)}{3\alpha}\left[1 - (1 + (\frac{3}{2-\alpha}\Delta z))^\alpha\right] & \text{if } \Delta z > -\frac{2-\alpha}{3} \end{cases} \quad (6.124)$$

where prime means differentiation with respect to s and

$$\Delta z = \min(z(t - \theta) - z, -\frac{2 - \alpha}{3}). \quad (6.125)$$

6.6.1 Linear theory

Assuming $\Delta z > -(2 - \alpha)/3$, the linearized problem for $z = 0$ is

$$z'' + az' + z + b(z(s - \theta) - z) = 0. \quad (6.126)$$

which is identical to Eq. (6.109). A specific lobe of the stability diagram has been determined experimentally. The parameters have been evaluated experimentally and are given by: $m = 10$ kg, $\omega_0 = 578.79 \ s^{-1}$, $\zeta = 0.0135$, and $F_c(d) = Kwd^\alpha = 553d^{0.41}$ for $w = 0.25$ mm. This gives $a = 0.027$ and $b \simeq 0.12$ mm^{-1} $\times w$ where w is measured in mm. The angular frequency Ω is increased between 840 rpm and 925 rpm (between 14 rot/s and 15.42 rot/s). Because $\theta = \omega_0\tau = \omega_0/\Omega$, $\Omega_\theta = 2\pi/\theta = 2\pi\Omega/\omega_0$ and we find that Ω_θ changes between 0.152 and 0.167. The stability diagram in terms of b and Ω_θ is shown in Figure 6.12.

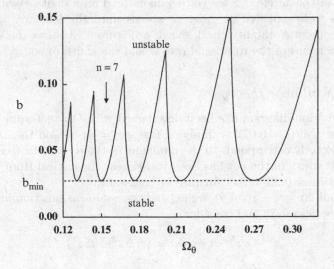

Figure 6.12: Blow-up of the stability diagram. The experimental value of Ω_θ corresponds to the minimum of the $n = 7$ lobe.

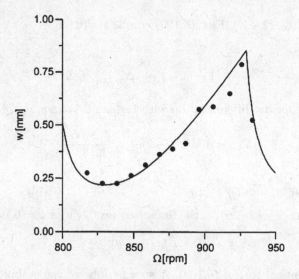

Figure 6.13: The line represents the analytical stability boundary for $n = 7$. Dots are the experimental estimates (from Kalmár-Nagy [121]).

The experiments have been done as follows. All cutting measurements were made with the same feed rate $sp = 508$ μm/s. Beginning with a small stable cut at a fixed speed, the width of the cut w was increased in 25 μm increments using the manual slide in a ramp and hold fashion. The experiment was stopped and recorded when self-sustained oscillations were observed in the accelerometer voltage monitored on a digital oscilloscope. The tests were repeated at different speeds until the lobe was mapped. The experimental stability chart shown in Figure 6.13 shows the stability domain in terms of the rotational rate Ω and the width of cut w.

6.6.2 Nonlinear theory

The bifurcation diagram has been determined with $\Omega = 836$ rpm $= 13.93$ rot s^{-1} ($\tau = 1/\Omega = 0.072$ s) implying $\theta \equiv \omega_0\tau = 41.67$ and $\Omega_\theta \equiv 2\pi/\theta = 0.15$. Ω_θ exactly corresponds to the minimum of the $n = 7$ lobe (see Figure 6.12). The width of the cut has been varied near the critical Hopf bifurcation point ($w = 0.23$) either forward or backward.

For small $\Delta z = z - z(s - \theta)$, we expand the nonlinear function in (6.124) in power series of Δz and consider

$$z'' + az' + z + b\Delta z = bc(\Delta z^2 - \Delta z^3), \qquad (6.127)$$

where

$$c = \frac{3(1 - \alpha)}{2(2 - \alpha)} = 0.56. \qquad (6.128)$$

As in the experiment, we concentrate on the minimum of the $n = 7$ stability lobe where the Hopf bifurcation point is located at $b = b_{min} = a(1 + a/2)$. In [120], the bifurcation analysis has been done for an arbitrary a and then the final result has been evaluated for low values of a. Here, we immediately take advantage of the small value of a in order to simplify the perturbation analysis. Specifically, we note that b_{min} is $O(a)$ small and seek a solution of the form

$$b = ab_1 + \dots \tag{6.129}$$

$$z = z_0(s, \zeta) + az_1(s, \zeta) + \dots \tag{6.130}$$

$$\theta = \theta_0 + a\theta_1 + \dots, \tag{6.131}$$

where $\zeta = as$ is a slow time variable. θ_0 is defined as the a small leading approximation of $\theta = \omega_0 \tau$ that corresponds to the minimum of the $n = 7$ stability lobe (see (6.120)). It is defined by

$$\theta_0 = -\pi/2 + 14\pi \simeq 42.41. \tag{6.132}$$

Introducing (6.129)–(6.132) into Eq. (6.127) and equating the coefficients of each power of a leads to a sequence of linear problems for $z_0, z_1, \dots$. The first two problems are

$$z_{0ss} + z_0 = 0, \tag{6.133}$$

$$z_{1ss} + z_1 = -z_{0s} - 2z_{0s\zeta} - b_1 \Delta z_0 + b_1 c(\Delta z_0^2 - \Delta z_0^3), \tag{6.134}$$

where $\Delta z_0 = z_0 - z_0(s - \theta)$. The solution of Eq. (6.133) is

$$z_0 = A(\zeta) \exp(is) + c.c., \tag{6.135}$$

where A is an unknown amplitude. The solvability condition for Eq. (6.134) then leads to the following equation for A,

$$2iA' = -iA - b_1(1 - i)A - 6b_1 c(1 - i)A^2 A^*. \tag{6.136}$$

Introducing the decomposition $A = R \exp(i\phi)$, the real and imaginary parts of Eq. (6.136) give

$$2R' = -R + b_1 R + 6b_1 cR^3, \tag{6.137}$$

$$2R\phi' = b_1 R + 6b_1 cR^3. \tag{6.138}$$

From Eq. (6.137), we determine the amplitude of the oscillations as

$$R^2 = \frac{1 - b_1}{6b_1 c} \geq 0 \tag{6.139}$$

and Eq. (6.138) provides the correction of the frequency of the oscillations

$$\phi' = \frac{b_1}{2} + 3b_1 cR^2 = \frac{1}{2}. \tag{6.140}$$

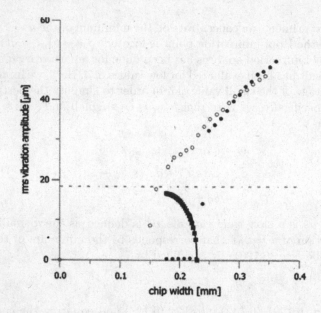

Figure 6.14: Experimental bifurcation diagram obtained by varying w forward (full circles) and backward (open circles). Squares are numerical simulations and the line is obtained analytically from a local analysis near the Hopf bifurcation point. The horizontal broken line is an estimate of the critical amplitude where the tool leaves the workpiece (from Kalmár-Nagy Ph.D. thesis with permission [121]).

The Hopf bifurcation is subcritical because the inequality in (6.139) implies

$$b_1 < 1. \tag{6.141}$$

In summary, we have found that the bifurcation is subcritical as suggested by the experiments. See Fig. 6.14. In contrast to Johnson and Moon's problem, the transition through the Hopf bifurcation point is not smooth (hard) and leads to large amplitude oscillations as soon as the Hopf bifurcation point is passed.

7
Lasers

Lasers are ubiquitous today, mostly in the form of semiconductor lasers (SLs, also known as laser diodes), which are characterized by smallness in size, weight, cost of production, and power requirements. Optical networks depend on SLs for generation, amplification, and distribution of the light that transmits voice, video, and data. They also are the lasers used in our everyday activities (CD player, laser printer, and barcode reading at the supermarket). However, an unfortunate property of these devices is their high susceptibility to unavoidable optical feedback (OFB), such as reflection from any optical element of the system surrounding the laser. Even tiny amounts of OFB (less than 0.01%) can cause the laser to enter a state of erratic pulsating instabilities and irregular chaotic transitions.

Systematic experimental investigations started in the early 1970s. If the external cavity is of the order of one meter, noise peaks appear at GHz frequencies and are referred to as "high-frequency noise". Moreover, "low-frequency" noise dominates at frequencies less than 100 MHz and appears to be proportional to the external cavity length. However, many effects resulting from OFB depend on the experimental conditions (see [70] for an excellent review). The interest of OFB arises from the rich phenomenology observed, ranging from multistability, bursting, intermittency, irregular intensity drops (low-frequency fluctuations or LFFs), and transition to developed chaos (coherence collapse). A complete understanding of the physical mechanisms as the basis of such complex behavior is, however, still missing. In particular, the origin of the LFF regime (stochastic, deterministic, or both) has been under debate since the very first observations and yet this puzzling problem has not been solved. A widely used theoretical

T. Erneux, *Applied Delay Differential Equations*, Surveys and Tutorials in the Applied Mathematical Sciences 3, DOI 10.1007/978-0-387-74372-1_7, © Springer Science+Business Media, LLC 2009

description of the system is the Lang–Kobayashi (LK) model [136] introduced in 1980 in an effort to provide a simplified but effective analysis of a SL optically coupled with a distant reflector. Significant progress has been made by comparing experimental and numerical simulations of the LK equations. These equations consist of three coupled equations for the amplitude and the phase of the laser field in the cavity, and the carrier density. However, some specific feedback configurations allow is to ignore the phase of the laser field and we first consider these systems.

We determine bifurcation diagrams of periodic solutions by massively using solvability conditions. These conditions are required to guarantee that the periodic solution is bounded at each order of the perturbation analysis. As we demonstrate, they are powerful tools to determine unknown quantities such as the amplitude or the frequency correction of the solution.

7.1   Optoelectronic feedback

A feedback system where the intensity of the field and the carrier density are the only dependent variables can be realized if the intensity of the laser field is detected electronically, amplified, and then reinjected into the pumping current of the laser.

An example of a semiconductor laser system and its optoelectronic feedback is sketched in Figure 7.1. Part of the laser output is detected with a

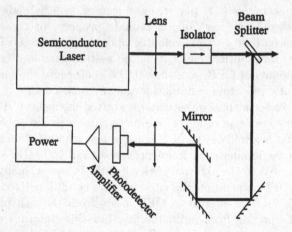

Figure 7.1: Semiconductor laser subject to an optoelectronic feedback. Part of the output light is injected into a photodetector connected to the pump. The feedback delay is controlled by changing the length of the optical path (full line). From Pieroux et al. [189] copyright © 2000 Society for Industrial and Applied Mathematics. Reprinted with permission. All rights reserved.

high-speed photodetector. The detector photocurrent component is selected with a T bias, amplified, and added to the DC pump current [203]. The laser rate equations modeling this system are given by

$$\frac{dI}{dt} = 2NI, \tag{7.1}$$

$$T\frac{dN}{dt} = P + \eta I(t - \tau) - N - (1 + 2N)I, \tag{7.2}$$

where I and N represent the intensity of the laser field and the electronic carrier density, respectively. Time t is measured in units of the photon lifetime τ_p $(t \equiv t'/\tau_p)$. $T \equiv \tau_n/\tau_p$ is the ratio of the carrier lifetime τ_n and the cavity lifetime τ_p; $\tau \equiv \tau'/\tau_p$ is the delay of the feedback where $\tau' = 2L/c$ is the cavity transit time (L is the distance laser-mirror and c is the speed of light); $P \sim J/J_{th} - 1$ is the pumping current above threshold. These equations are the well-known SL rate equations rewritten in dimensionless form. Only P is replaced by $P + \eta|Y(t - \tau)|^2$ to take into account the effect of the DC coupled optoelectronic feedback. Typical values of the parameters are $\tau_p \sim 2$ ps, $\tau_n \sim 2$ ns, and $\tau \sim 1 - 10$ ns which mean that $T \sim 10^3$ and $\tau \sim 10^3$ are relatively large parameters.

The large T multiplying the left-hand side of Eq.(7.2) is a source of numerical difficulties but can be removed by a change of variables. We introduce x, y, and s defined as

$$I = P(1 + y), \quad N = \frac{\omega_r}{2}x, \quad s = \omega_r t, \tag{7.3}$$

where

$$\omega_r \equiv \sqrt{2P/T} \tag{7.4}$$

is known as the laser relaxation oscillation (RO) frequency. Inserting (7.3) into (7.1) and (7.2), we obtain

$$x' = -y + \eta\left(1 + y(s - \theta)\right) - \varepsilon x\left[1 + \frac{2P}{1 + 2P}y\right], \tag{7.5}$$

$$y' = (1 + y)x, \tag{7.6}$$

where prime now means differentiation with respect to time s. The new parameters ε and θ are defined by

$$\varepsilon \equiv \omega_r\frac{(1 + 2P)}{2P} \quad \text{and} \quad \theta \equiv \omega_r\tau. \tag{7.7}$$

The large T parameter is now located in ε which is small because ω_r is proportional to $T^{-1/2}$ and $T \sim 10^3$. Because τ is of the same size as T, the rescaled delay θ is proportional to $T^{1/2}$ and remains a relatively large parameter.

7.2   Delayed incoherent feedback

For a laser subject to optoelectronic feedback, the bandwidth of the electronics is required to be very large and flat because the chaotic dynamics of the SL can span tens of GHz. If the bandwidth of the electronics does not match the speed of the optical intensity fluctuations, the response of the laser system will be dominated by this bandwidth limitation. An alternative way to investigate this specific delayed feedback problem is to consider a SL subject to incoherent optical feedback [180, 188]. The output field of a SL is reinjected into the laser cavity after rotation of the polarization to the orthogonal state, providing a delayed feedback that affects only the carriers. The advantage is that the feedback remains purely optical. This system has been experimentally studied in [205]. The laser device is described by the following equations for the intensity of the laser field I and the carriers N,

$$\frac{dI}{dt} = 2NI, \tag{7.8}$$

$$T\frac{dN}{dt} = P - N - (1 + 2N)\left[I + \eta I(t - \tau)\right], \tag{7.9}$$

where the last term in Eq. (7.9) represents the reinjected orthogonal polarization intensity. With the new variables previously introduced in (7.3), Eqs. (7.8) and (7.9) take the form

$$x' = -y - \eta\left(1 + y(s - \theta)\right)$$
$$-\varepsilon x\left[1 + \frac{2P}{1 + 2P}\left(y + \eta(y + y(s - \theta))\right)\right], \tag{7.10}$$

$$y' = (1 + y)x. \tag{7.11}$$

If η is $O(\varepsilon)$ small, we may neglect the $\varepsilon\eta$ term in Eq. (7.10) and we find Eq. (7.5). Thus the optoelectronic feedback and incoherent feedback problems are described by the same equations if the feedback rate is sufficiently low. In the next subsections, we analyze these equations for low and moderate values of the feedback rate.

7.2.1   Small feedback rate

We wish to analyze Eqs. (7.10) and (7.11) in the limit ε small assuming

$$\eta = \varepsilon C, \tag{7.12}$$

where $C = O(1)$. Specifically, we seek a solution of Eqs. (7.10) and (7.11) of the form

$$x = x_0(s) + \varepsilon x_1(s) + \dots \tag{7.13}$$

$$y = y_0(s) + \varepsilon y_1(s) + \dots. \tag{7.14}$$

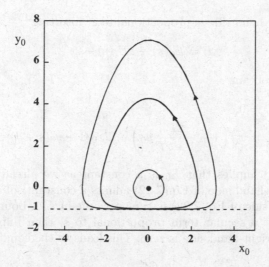

Figure 7.2: One-parameter family of periodic solutions. Different periodic orbits surrounding the origin are shown in the phase plane (x_0, y_0). They are all bounded below by the invariant line $y = -1$.

After introducing (7.13) and (7.14) into Eqs. (7.10) and (7.11), we find that (x_0, y_0) satisfies

$$x_0' = -y_0, \tag{7.15}$$
$$y_0' = x_0(1 + y_0). \tag{7.16}$$

These equations admit a one-parameter family of periodic solutions. See Figure 7.2. The first integral is given by

$$E_0 = \frac{x_0^2}{2} + y_0 - \ln(1 + y_0), \tag{7.17}$$

where E_0 is a constant of integration that depends on the initial conditions $(x_0(0), y_0(0))$. We denote by $(x_0, y_0) = (X, Y)$ the solution of period P. Because we ignore how this period depends on the parameters C and θ, we investigate the problem for (x_1, y_1) and formulate a solvability condition. This condition can be derived in the following way. Suggested by (7.17), we introduce the energy $E = E(x, y)$ defined as

$$E \equiv \frac{x^2}{2} + y - \ln(1 + y). \tag{7.18}$$

Differentiating E with respect to s and using Eqs. (7.10) and (7.11), we find the following equation for E,

$$E' = \varepsilon \left[x^2 \left(1 + \frac{2P}{1 + 2P} y \right) + Cx \left(1 + y(s - \theta) \right) \right], \tag{7.19}$$

where the right-hand side is proportional to ε. Inserting

$$E = E_0(s) + \varepsilon E_1(s) + \dots \tag{7.20}$$

into (7.19), we obtain

$$E_0' = 0, \tag{7.21}$$

$$E_1' = x_0^2 \left(1 + \frac{2P}{1 + 2P} y_0\right) + C x_0 \left[1 + y_0(s - \theta)\right]. \tag{7.22}$$

Equation (7.21) implies that E_0 is a constant as we already knew from (7.17). The left-hand side of Eq.(7.22) admits a constant solution whereas the right-hand side of Eq.(7.22) is time-periodic in s. A bounded solution for E_1, without a secular term proportional to s, then implies that the average of the right-hand side is zero. This leads to the condition

$$\int_0^P X^2 ds + C \int_0^P XY(s - \theta) ds = 0, \tag{7.23}$$

where we have used the fact that[1]

$$\int_0^P x_0^2 y_0 ds = 0 \quad \text{and} \quad \int_0^P x_0 ds = 0. \tag{7.24}$$

Equation (7.23) is the bifurcation equation for the P-periodic solutions relating the amplitude of $(x_0, y_0) = (X, Y)$ to the parameters C and θ. Multiplying by ε, Eq. (7.23) is rewritten in terms of the original parameters ε and η,

$$\varepsilon \int_0^P X^2 ds + \eta \int_0^P XY(s - \theta) ds = 0. \tag{7.25}$$

$\varepsilon \ll 1$ measures the damping of the relaxation oscillations and η is the feedback parameter. If we neglect the damping ($\varepsilon = 0$), (7.25) reduces to the integral

$$\int_0^P XY(s - \theta) ds = 0 \tag{7.26}$$

which is verified if

$$\theta = n\frac{P}{2} \quad (n = 0, 1, 2, \dots). \tag{7.27}$$

The different branches of time-periodic solutions satisfying (7.27) are shown in Figure 7.3. Equation (7.27) can be explained as follows. We may redefine time s, without loss of generality, such that $X(s)$ is an even function of s and so $s = 0$ corresponds to $Y(0) = \min(Y)$ and $X(0) = 0$ (see Figure 7.4).

---

[1]Using (7.15), $y_0 ds = -dx_0$. The resulting integrals on x_0 are zero because we integrate over a complete period.

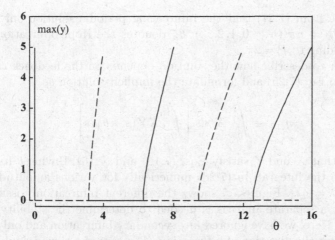

Figure 7.3: Successive branching of limit-cycle solutions as $\eta \to 0^+$. Full and broken lines represent stable and unstable solutions, respectively.

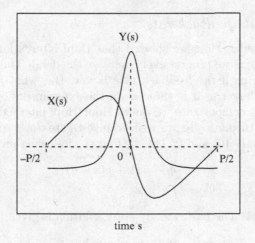

Figure 7.4: One-periodic solution of the conservative system. $X(s)$ and $Y(s)$ are an odd and even function of s, respectively.

If $X(s)$ is an even function of s, $Y = -X'$ is an odd function of s on the interval $-P/2 < s < P/2$. Because $Y(s - nP/2)$ is also an odd function of s, (7.26) is an integral of an odd-periodic function over one period and is equal to zero.

If $n = 0$, $\theta = 0$ for all amplitudes (vertical bifurcation) and the role of damping (i.e., ε nonzero) will be important. But if $n \neq 0$, (7.27) is an amplitude equation because P is a function of the amplitude. Because P is a monotonic increasing function of the amplitude of (X, Y) $(2\pi \leq P < \infty)$, we deduce from (7.27) that the bifurcating periodic solution only exists for $\theta > \theta_n = n\pi$ $(n = 0, 1, 2, ...)$. θ_n denotes the Hopf bifurcation point corresponding to $P = 2\pi$.

In order to describe how the solution depends on the feedback rate, we go back to Eq. (7.25) and formulate the implicit solution as

$$\eta = -\varepsilon \int_0^P X^2 ds \left[\int_0^P XY(s-\theta)ds \right]^{-1}. \tag{7.28}$$

Recall that X and Y satisfy Eqs. (7.15) and (7.16). Having θ fixed, we may solve the integrals in (7.28) numerically for various amplitudes and obtain[2] $C = \eta/\varepsilon$. Figure 7.5 shows the different bifurcation possibilities. Note that a separate analysis is needed to determine the stability of the branches. Here, we have ignored any secondary bifurcation and only taken into account the direction of bifurcation. The case of a period-doubling bifurcation is analyzed in detail in [188]. Figure 7.6 illustrates the coexistence of multiple stable periodic solutions.

7.2.2   Moderate feedback rate

In the previous subsection, we showed that Hopf bifurcation branches sequentially appear as we progressively increase the delay. This is the typical bifurcation diagram if the feedback rate is low. But what happens if the feedback rate is larger as it is the case in the experiments by Lin and Liu [142]? As we now demonstrate, resonant Hopf–Hopf interactions are possible [41]–[43]. Mathematically, we wish to investigate the case $\eta = O(1)$ but $\varepsilon \ll 1$. From Eqs. (7.10) and (7.11), we find the following problem if $\varepsilon = 0$,

$$x' = -y - \eta \left(1 + y(s-\theta) \right), \tag{7.29}$$
$$y' = (1+y)x, \tag{7.30}$$

where $-1 < \eta < 1$. The basic steady state solution is given by

$$x_0 = 0 \quad \text{and} \quad y_0 = -\eta(1+\eta)^{-1}. \tag{7.31}$$

From the linearized equations, we then find the following characteristic equation for the growth rate σ,

$$\sigma^2 + 1 + \eta \exp(-\sigma\theta) = 0. \tag{7.32}$$

By analyzing the implicit function $\eta = -(\sigma^2 + 1) \exp(\sigma\theta)$ (θ fixed), we note that there exists a real root for $-1 < \eta < 0$ that is always negative. We

---

[2]As the amplitude of X and Y increases, the periodic solution is pulsating in time and its numerical determination becomes harder because of the high accuracy needed when Y is (exponentially) close to -1. To remove part of this difficulty, we integrate the laser equations in term of X and $Z = -\ln(1+Y)$ when Y is close to -1.

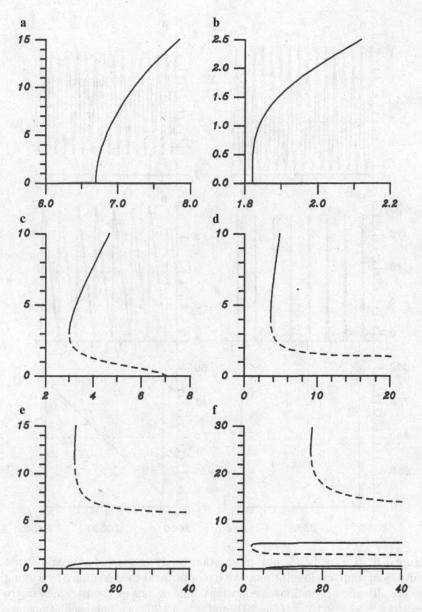

Figure 7.5: The maximum of X is shown as a function of the scaled feedback rate $C = \eta/\varepsilon$ for increasing values of the delay θ. Stable and unstable solutions are represented by full and dotted lines, respectively. (a) $\theta = 0.15$ supercritical Hopf bifurcation; (b) $\theta = \theta_c \simeq 2.26$ vertical Hopf bifurcation; (c) $\theta = 3$ subcritical Hopf bifurcation; (d) $\theta = 3.3$ isolated branch of periodic solutions; (e) $\theta = 6.45$, coexistence of bifurcating and isolated branches; (f) $\theta = 12.75$, coexistence of a bifurcating branch and two isolated branches (reprinted Fig. 4 with permission from Pieroux et al. [188] copyright 1994 American Physical Society).

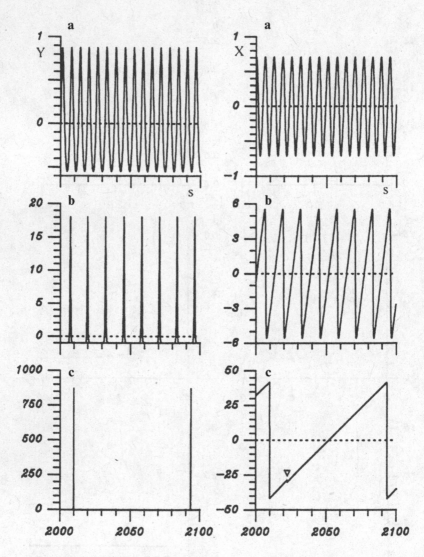

Figure 7.6: If $\theta = 12.75$, our bifurcation analysis predicts three distinct branches of limit-cycle solutions. We have obtained these regimes by solving numerically the original laser equations. The values of the parameters are $\varepsilon = 10^{-3}$, $\eta = 2 \times 10^{-2}$ ($C = 20$), and $\theta = 12.75$. (a) The oscillations are nearly harmonic and result from a Hopf bifurcation. The period is close to 2π, (b) The oscillations are pulsating and the period is equal to the delay θ, (c) The oscillations are of high intensity and strongly pulsating. The period is much larger than θ ($\simeq 6.5\theta$). The arrow in the figure indicates the effect of the function $y(s - \theta)$ which is pulsating after the pulse of $y(s)$ (reprinted Fig. 6 with permission from Pieroux et al. [188] copyright 1994 American Physical Society).

thus concentrate on the complex roots and examine the case of the Hopf bifurcations. Introducing $\sigma = i\omega$ into Eq. (7.32), we obtain the following two conditions for a Hopf bifurcation

$$(1 + \eta)\omega^2 = 1 + \eta\cos(\omega\theta), \tag{7.33}$$
$$\sin(\omega\theta) = 0. \tag{7.34}$$

We obtain the following two possibilities

(1) $\theta = 2n\pi$ and $\omega = 1$ $(n = 1, 2, ...)$ $\hspace{2cm}$ (7.35)

(2) $\theta = (1 + 2m)\pi$ and $\omega = (1 + 2m)\pi\sqrt{\dfrac{1 + \eta}{1 - \eta}}$ $(m = 0, 1, 2...)$. $\hspace{0.3cm}$ (7.36)

The Hopf bifurcation lines are shown in Figure 7.7. The stable regions are indicated and have been found by determining the sign of the real part of σ near each Hopf bifurcation line. Note that the points of intersection are double Hopf points with two distinct frequencies. See the following table. Because $\omega_1\theta = 2n\pi$ and $\omega_2\theta = (1 + 2m)\pi$, the ratio of the two frequencies at the double Hopf points is a ratio of two integers: $\omega_1/\omega_2 = n/(1 + 2m)$. This means that near these points, a secondary bifurcation to a periodic solution is likely to appear. An example of a secondary bifurcation is shown in Figure 7.8.

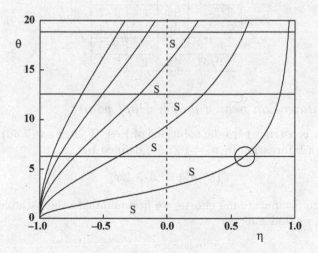

Figure 7.7: Hopf bifurcation lines in the θ versus η plane. The regions marked by S correspond to a stable steady state. The different bifurcation lines intersect at double Hopf points (two distinct frequencies). The bifurcation diagram near the point encircled is examined in detail.

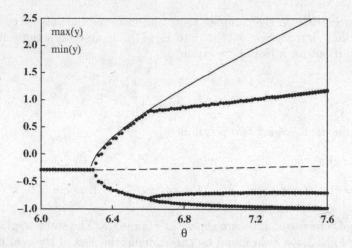

Figure 7.8: Secondary bifurcation. Bifurcation diagram of the stable solutions for $\eta = 0.4 < \eta^* = 0.6$ and for decreasing values of θ ($6 < \theta < 7.6$). The Hopf bifurcation at $\theta_1 = 2\pi \simeq 6.28$ is followed by a period-doubling bifurcation at $\theta_2 \simeq 6.62$. Note that the branch emerging at $\theta = \theta_1$ can be computed independently using the fact that its period P equals θ (full line).

η	θ	ω_1	ω_2
$3/5$	2π	1	$1/2$
$7/25$	4π	1	$3/4$
$-5/13$	2π	1	$2/3$
$-9/41$	4π	1	$4/5$

7.2.3   Bifurcation near a double Hopf point

We wish to construct periodic solutions of Eqs. (7.29) and (7.30) near the double Hopf bifurcation $(\eta, \theta) = (\eta^*, \theta^*)$ defined by

$$(\eta^*, \theta^*) = (3/5, 2\pi). \qquad (7.37)$$

To minimize computational efforts, we first eliminate the variable x. From Eq. (7.30), we find x as

$$x = \frac{y'}{1+y} = (\ln(1+y))'. \qquad (7.38)$$

Inserting (7.38) into Eq. (7.29), we obtain

$$\ln(1+y)'' = -y - \eta\,(1 + y(s - \theta)) \qquad (7.39)$$

which is a second order differential equation for y only. We next introduce the deviation from the steady state (7.31). Specifically, we introduce Y defined by

$$Y \equiv -1 + \frac{1+y}{1+y_0},\qquad(7.40)$$

where y_0 is given by (7.31). From (7.39), we obtain the following equation for Y,

$$\ln(1+Y)'' = -\frac{1}{1+\eta}(-Y + \eta Y(s-\theta)).\qquad(7.41)$$

Finally, we introduce the new time variable $S \equiv s\pi/\theta$ so that $Y(s-\theta) = Y(S-\pi)$ exhibits a fixed delay. The resulting equation is given by

$$(1+\eta)\ln(1+Y)'' = -(\frac{\theta}{\pi})^2 \left[Y + \eta Y(S-\pi)\right],\qquad(7.42)$$

where prime means differentiation with respect to time S. Equation (7.42) is ready for the perturbation analysis. We first introduce a small parameter $0 < \nu \ll 1$ defined as

$$\eta = \eta^* + \nu\eta_1,\qquad(7.43)$$

where $\eta_1 = \pm 1$. We next expand the parameter θ as

$$\theta = \theta^*(1 + \nu\delta_1 + \nu^2\delta_2 + ...)\qquad(7.44)$$

and seek a 2π-periodic solution of the form

$$Y = \nu Y_1(T) + \nu^2 Y_2(T) + ...,\qquad(7.45)$$

where the new time T is defined by

$$T \equiv (1 + \nu\sigma_1 + \nu^2\sigma_2 + ...)S.\qquad(7.46)$$

Inserting (7.43)–(7.46) into Eq. (7.42) and equating to zero the coefficients of each power of ν lead to a sequence of linear problems for Y_1, Y_2, ... The first two problems are

$$LY_1 \equiv \frac{8}{5}Y_1'' + 4\left[Y_1 + \frac{3}{5}Y_1(T-\pi)\right] = 0,\qquad(7.47)$$

$$LY_2 = -\eta_1 Y_1'' - \frac{16}{5}\sigma Y_1'' + \frac{4}{5}(Y_1^2)'' - 8\delta_1\left[Y_1 + \frac{3}{5}Y_1(T-\pi)\right]$$
$$- 4\eta_1 Y_1(T-\pi) + \frac{12\pi}{5}\sigma Y_1'(T-\pi),\qquad(7.48)$$

where prime now means differentiation with respect to time T. The solution of Eq. (7.47) is

$$Y_1 = A_1\exp(iT) + c.c. + A_2\exp(2iT) + c.c.,\qquad(7.49)$$

where A_1 and A_2 are two unknown amplitudes. Note that two periodic functions [namely, $\exp(iT)$ and $\exp(2iT)$] contribute to the solution because of the double Hopf bifurcation point. This implies that we need to apply two solvability conditions with respect to each periodic function. Because the left-hand side is a single linear second-order equation, these conditions are easy to formulate. We need to expand the right-hand side of Eq. (7.48) and set the coefficients of $\exp(iT)$ and $\exp(2iT)$ equal to zero. This leads to two coupled equations for A_1 and A_2 given by

$$[(4 - 3\pi i)4\sigma_1 + (25\eta_1 - 16\delta_1)] A_1 - 8A_1^* A_2 = 0, \qquad (7.50)$$

$$[(8 + 3\pi i)\sigma_1 - 8\delta_1 A_2] A_2 - 2A_1^2 = 0. \qquad (7.51)$$

We now investigate these conditions. The cases $\sigma_1 = 0$ and $\sigma_1 \neq 0$ must be treated separately.

Secondary bifurcation

In terms of $A_j = R_j \exp(i\phi_j)$, the solution of Eqs. (7.50) and (7.51) with $\sigma_1 = 0$ satisfies the two conditions

$$R_2 = -\frac{R_1^2}{4\delta_1} \exp(-i\Phi) \quad \text{and} \quad R_1 \left[\delta_1(-25\eta_1 + 16\delta_1) - 2R_1^2\right] = 0, \quad (7.52)$$

where $\Phi = \phi_2 - 2\phi_1$. The first solution is the trivial solution

$$(1) : R_1 = R_2 = 0. \qquad (7.53)$$

The second solution verifies $R_1 \neq 0$ and $R_2 \neq 0$ and is given by

$$(2) \; : \; R_1^2 = \frac{\delta_1}{2}(-25\eta_1 + 16\delta_1) > 0, \qquad (7.54)$$

$$R_2 = -\frac{1}{8}(-25\eta_1 + 16\delta_1)\cos(\Phi) > 0, \qquad (7.55)$$

where $\Phi = 0$ or π. A third solution is possible if $\delta_1 = 0$ and has the form

$$(3) : \delta_1 = 0, \quad R_1 = 0 \quad \text{but } R_2 \text{ arbitrary.} \qquad (7.56)$$

It corresponds to a vertical Hopf bifurcation. The connection between Solution (2) and Solution (3) appears at a secondary bifurcation with critical amplitude

$$R_2 = \pm\frac{25}{8}\eta_1 > 0. \qquad (7.57)$$

See Figure 7.9. But because $\delta_1 = 0$ and R_2 is unknown for Solution (3), we need to investigate the higher-order problem. The solution of Eq. (7.48) with $Y_1 = A_2 \exp(2iS) + c.c.$ and $\delta_1 = 0$ is

$$Y_2 = B_1 \exp(iT) + c.c. + B_2 \exp(2iT) + c.c. + \frac{2}{3}A_2^2 \exp(4iT) + c.c., \quad (7.58)$$

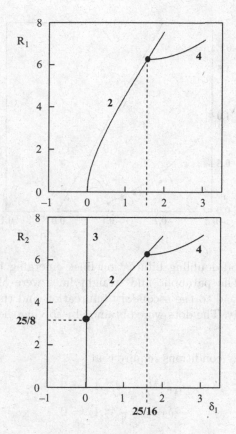

Figure 7.9: Bifurcation diagram of R_1 and R_2 as a function of δ_1. The labels 2, 3, and 4 refer to the different solution branches determined analytically. The bifurcation at $R_2 = 25/8$ marks a period-doubling bifurcation. As δ_1 progressively increases from zero, R_1 progressively increases but the period remains constant. At the point where $R_1 = R_2$ ($\delta_1 = 25/16$), the period-2 solution experiences a bifurcation to a new period-2 solution where the period is now a function of δ_1.

where B_1 and B_2 are two new unknown amplitudes. The problem for Y_3 with $\delta_1 = \sigma_1 = \sigma_2 = 0$ is

$$LY_3 = -\eta_1 Y_2'' + \frac{8}{5}(Y_2 Y_1' + Y_1 Y_2' - Y_1^2 Y_1')'$$

$$- 8\delta_2(Y_1 + \frac{3}{5}Y_1(S - \pi)) - 4\eta_1 Y_2(S - \pi). \tag{7.59}$$

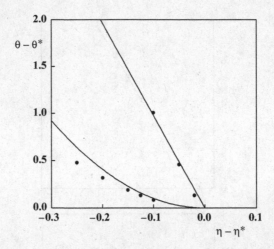

Figure 7.10: Period-doubling bifurcation lines emerging from the double 1:2 Hopf point. The parabolic and straight lines were obtained analytically and correspond to the secondary bifurcation and the tertiary bifurcation, respectively. The dots were obtained by solving the laser equations numerically.

The two solvability conditions require that

$$\eta_1 25 B_1 - 8 B_1^* A_2 = 0, \tag{7.60}$$

$$-2\delta_2 A_2 + \frac{1}{3} A_2^2 \overline{A_2} = 0 \tag{7.61}$$

implying that either

$$B_1 = 0 \quad \text{and} \quad R_2^2 = 6\delta_2 > 0 \tag{7.62}$$

or $B_1 \neq 0$ and

$$R_2^2 = 6\delta_2 = (\frac{25}{8}\eta_1)^2. \tag{7.63}$$

The expression (7.63) locates the secondary bifurcation in parameter space (θ, η). Using (7.43) and (7.44), we find

$$\frac{\theta - \theta^*}{\theta^*} = \frac{1}{6} \left(\frac{25}{8}\right)^2 (\eta - \eta^*)^2. \tag{7.64}$$

The parabolic function is shown in Figure 7.10.

Tertiary bifurcation

We next seek a solution of Eqs. (7.50) and (7.51) of the form $A_j = R_j \exp(i\phi_j)$ with $\sigma_1 \neq 0$. The first nontrivial solution is $R_1 = 0$ and $R_2 \neq 0$. However, this solution is possible only if $\delta_1 = \sigma_1 = 0$ which is a case that was previously studied. The second solution verifies $R_1 \neq 0$ and $R_2 \neq 0$. From the real and imaginary parts, we find the following four equations

$$16\sigma_1 + 25\eta_1 - 16\delta_1 - 8R_2\cos(\Phi) = 0, \tag{7.65}$$
$$-\sigma_1 12\pi - 8R_2\sin(\Phi) = 0, \tag{7.66}$$
$$8\sigma_1 R_2 - 8\delta_1 R_2 - 2R_1^2 = 0, \tag{7.67}$$
$$3\pi\sigma_1 R_2 - 2R_1^2\sin(\Phi) = 0. \tag{7.68}$$

Adding $R_2\times$(7.66) and $4\times$(7.68) leads to the condition

$$8(R_1^2 - R_2^2)\sin(\Phi) = 0. \tag{7.69}$$

A first solution of Eq. (7.69) is $\sin(\Phi) = 0$ meaning $\sigma_1 = 0$ from (7.66) or (7.68). This case was studied in the previous subsection. A second possibility verifying condition (7.69) is

$$(4) : R_1 = R_2 = R. \tag{7.70}$$

Equations (7.65)-(7.68) then simplify as

$$16\sigma_1 + 25\eta_1 - 16\delta_1 - 8R\cos(\Phi) = 0, \tag{7.71}$$
$$-\sigma_1 12\pi - 8R\sin(\Phi) = 0, \tag{7.72}$$
$$8\sigma_1 - 8\delta_1 - 2R\cos(\Phi) = 0, \tag{7.73}$$
$$3\pi\sigma_1 + 2R\sin(\Phi) = 0. \tag{7.74}$$

Eliminating $R\cos(\phi)$ from Eqs. (7.71) and (7.73), we obtain an expression for σ_1 given by

$$\sigma_1 = \frac{25}{16}\eta_1 + \delta_1. \tag{7.75}$$

Note from (7.54) and (7.55) with $\Phi = \pi$, $\delta_1 > 0$, and $\eta_1 < 0$ that the point where $R_1 = R_2$ of Solution (2) appears at

$$\delta_1 = -\frac{25\eta_1}{16}, \tag{7.76}$$

which is exactly the point where σ_1 defined by (7.75) is zero. This point is a tertiary bifurcation point. In terms of θ, it is located at

$$\frac{\theta - \theta^*}{\theta^*} = -\frac{25}{16}(\eta - \eta^*). \tag{7.77}$$

The function is the straight line shown in Figure 7.10. Finally, from (7.72) and (7.73), we obtain $R\sin(\Phi) = -\sigma_1 3\pi/4$ and $R\cos(\Phi) = 4(\sigma_1 - \delta_1)$ which lead to

$$R = \sqrt{\frac{\sigma^2 9\pi^2}{16} + 16(\sigma_1 - \delta_1)^2} \tag{7.78}$$

which describes the amplitude of solution (7.70). See Figure 7.9.

In summary, we determined analytically three successive bifurcations. Their locations compare well with the numerical estimates obtained by simulating the original laser equations. See Fig. 7.11.

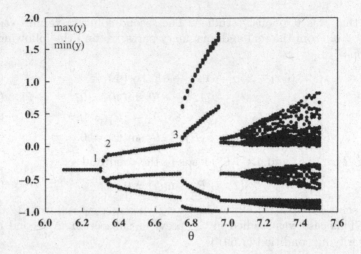

Figure 7.11: Cascading bifurcations. Numerical bifurcation diagram of the stable solutions obtained by decreasing θ from $\theta = 7.6$ to $\theta = 6.1$. The value of $\eta = 0.55 < \eta^* = 0.6$. The three first bifurcation points found numerically are located at: $\theta_1 = 2\pi \simeq 6.28$, $\theta_2 \simeq 6.31$ ($\theta_{2anal} \simeq 6.31$), and $\theta_3 \simeq 6.75$ ($\theta_{3anal} \simeq 6.77$).

7.3   Delayed coherent feedback

OFB from distant reflectors may have dramatic effects on the laser normal output. This is the case for SLs used in telecommunication, data transmission, and data storage technologies. Moreover, because the RO frequency is of the order of a few GHz, the laser is particularly sensitive to GHz signals, threatening reliable performance in optical communication systems. Careful isolation of the laser is possible but increases the complexity and cost of the laser system.

The response of the laser due to OFB is rich and varied. Typical experiments in the laboratory consider the case of an external mirror located at one to several meters from the laser [98]. See Fig. 7.12. The round-trip frequency of light $\nu_{EC} = \tau^{-1}$ is then some hundreds of MHz and is substantially lower than the GHz range relaxation oscillation frequency ν_{RO}. The output intensity is chaotic but is characterized by at least two distinct time-scales. Figure 7.13 shows an example recorded for injection currents close to the solitary laser threshold. Figure 7.13 (top) shows irregular fluctuations of the laser intensity on a time scale of microseconds, which is very slow compared with the RO period. Figure 7.13 (bottom) shows the same dynamics on faster time-scales indicating that indeed there is a faster dynamics ($\nu_{average} \simeq \nu_{RO}$) underlying the slow dynamics. Note in

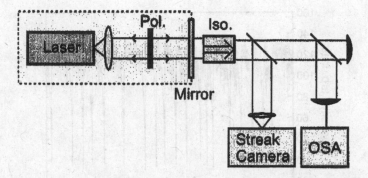

Figure 7.12: Typical experimental set-up. A temperature-stabilized laser diode is subject to delayed optical feedback from a semitransparent dielectric mirror. The laser beam is collimated using an aspheric lens, and feedback strength is controlled with a polarizer (Pol.). The optical isolator (Iso.) shields this external cavity configuration (rectangle in broken lines) from the detection branch. The light is analyzed using a single-shot streak camera and the optical spectrum is monitored with a grating spectrometer (reprinted Fig. 1 with permission from Heil et al. [87], copyright 2001 American Physical Society).

Figure 7.13 (top) the irregular intensity drops. This phenomenon is called Low Frequency Fluctuations or LFF.

However, in many practical applications such as fiber couplers or in compact discs, the external cavity is only a few centimeters long. The ratio of the two basic frequencies ν_{EC} and ν_{RO} is reversed and a different laser response is possible [87]. See Figure 7.14. We note that the intensity output is more regular than the one shown in Figure 7.13. The laser intensity shows a periodic emission of regular pulse packages separated by short intervals of very low intensity. The dynamics on the short time-scale is now dominated by the delay time.

Mathematically, we consider the idealized case of a single mode laser subject to a weak optical feedback so that multiple reflection can be ignored. With the injection of the delayed optical field [136], the laser equations for the amplitude of the field Y and the carrier density Z are given by

$$\frac{dY}{dt} = (1 + i\alpha)ZY + \eta \exp(-i\Omega_0\theta)Y(t - \theta), \tag{7.79}$$

$$T\frac{dZ}{dt} = P - Z - (1 + 2Z)|Y|^2, \tag{7.80}$$

where $t = t'/\tau_p$. P is the pump parameter above threshold ($P = O(1)$), $T \equiv \tau_c/\tau_p$, $\theta \equiv \tau_L/\tau_p$ are ratios of times, η is the feedback rate ($\eta \ll 1$), and $\Omega_0\theta \equiv \omega_0\tau$ is a phase called the feedback phase. The delay τ appears in the delayed field amplitude $Y(t - \tau)$ and in the phase factor $\exp(-i\omega_0\tau)$.

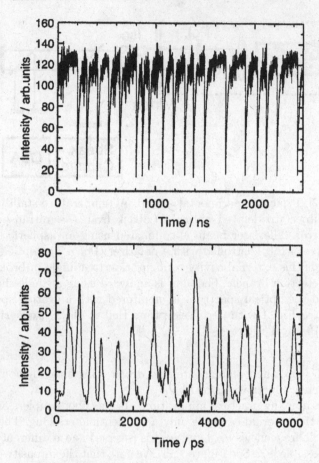

Figure 7.13: Intensity time series recorded for a laser operating close to its threshold. Top: Oscilloscope single-shot measurement, bandwidth 1 GHz. Bottom: streak camera single-shot measurement, bandwidth more than 50 GHz (from Heil et al. [88]).

Varying the position of the external mirror over one half optical wavelength (250–750 nm) corresponds to a variation in the phase $\omega_0 \tau_L$ of 2π.

The LK equations (7.79) and (7.80) have been studied analytically and numerically. Computer simulations have shown that they correctly describe the dominant effects observed experimentally. These include the occurrence of mode hopping [211, 171, 232], low-frequency fluctuations [170, 204], or the onset of "coherence collapse" [211, 179]. The LK equations admit simple solutions called the external cavity modes (ECMs). These modes are the reference solutions for all analytical or numerical bifurcation studies.

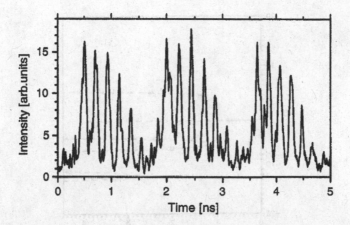

Figure 7.14: Streak camera measurements of the intensity time series of the laser operating in the short cavity regime. The injection curent is $I = 1.15I_{th,sol}$. The external cavity is 3.2 cm long corresponding to $\nu_{EC} = 4.7$ Ghz (reprinted Fig. 2 with permission from Heil et al. [87], copyright 2001 American Physical Society).

7.3.1 Single-mode solutions

A basic solution of Eqs. (7.79) and (7.80) is the single-frequency solution

$$Y = A \exp\left(i(\Omega - \Omega_0)t\right) \quad \text{and} \quad Z = B, \qquad (7.81)$$

where A, Ω, and B are constants. Equation (7.81) is called an external cavity mode or ECM. Substituting (7.81) into Eqs. (7.79) and (7.80) leads to three equations for A, B, and $\Delta \equiv \Omega\theta$ given by

$$B = -\eta \cos(\Delta), \qquad (7.82)$$

$$\Delta - \Omega_0\theta = -\eta\theta\left(\alpha\cos(\Delta) + \sin(\Delta)\right) \qquad (7.83)$$

$$A^2 = \frac{P + \eta\cos(\Delta)}{1 - 2\eta\cos(\Delta)} \geq 0. \qquad (7.84)$$

Δ is called the ECM frequency. It satisfies the transcendental equation (7.83) and the implicit solution is

$$\eta\theta = -\frac{\Delta - \Omega_0\theta}{\alpha\cos(\Delta) + \sin(\Delta)}. \qquad (7.85)$$

By continuously changing Δ from negative to positive values, we determine $\eta\theta$ from Eq. (7.85) and find several branches of solutions. See Figure 7.15. Except for the first mode that appears at $\eta = 0$, all other modes emerge by pairs from limit points and their number progressively increases with η.

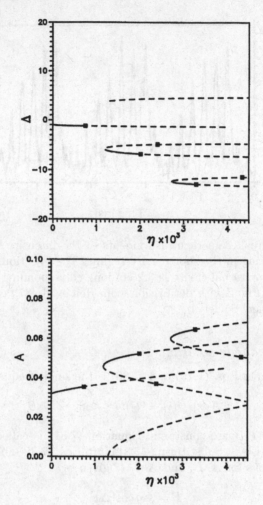

Figure 7.15: Bifurcation diagram of the ECMs. The upper and lower figures represent the ECM frequency and amplitude Δ and A, respectively. The values of the parameters are $P = 10^{-3}$, $T = \theta = 10^3$, $\alpha = 4$, and $\Omega_0\theta = -1$. Full and broken lines represent stable and unstable solutions, respectively. Black squares denote Hopf bifurcation points.

7.3.2   Two mode solutions and mode beating

Optical sources pulsating with high frequencies of several tens of gigahertz are required for a number of signal-processing applications. By the end of the 1990s, the group of Bernd Sartorius from the Heinrich-Hertz-Institut (HHI) in Berlin started to be interested in laser devices capable of generating tunable self-pulsations (SPs) with frequencies above 20 GHz. It was later discovered that Tager and Elenkrig [222] and Tager and Petermann

[223] already were concerned with this problem. Using the single-mode LK equations, they analyzed the possibility of a Hopf bifurcation to a high-frequency mode–antimode beating (MB) regime. They found that a short external cavity and a high feedback rate were necessary for this type of output. But the authors didn't give any clue on the stability of such high-frequency SPs. Could a stable mode and an unstable mode combine and produce a stable two-mode regime? Starting in 1999,[3] a series of workshops was organized at the WIAS by Klaus Schneider with the aim of attracting mathematicians and engineers and discuss these issues. In 2000, an asymptotic analysis of the LK equations based on the relative slow time-scale of the carriers compared to the photon lifetime showed that the high-frequency MB regimes belong to branches that connect isolated ECM branches (bridges) [57, 190]. Therefore, the LK equations may exhibit two types of Hopf bifurcations, namely, the bifurcation to RO oscillations or the bifurcation to MB regimes. How these bifurcations interact in parameter space was carefully investigated in [243]. In 2002, Sieber [213] proposed a detailed bifurcation analysis of the traveling wave laser equations emphasizing the domains of parameters where the high-frequency pulsations are possible. To achieve the required high feedback, Bauer et al. [11] from the HHI have attached to the passive short external cavity an active amplifier section. The carriers in the amplifier introduce an additional degree of freedom leading to a stabilization of the MB regime [13] as well as a complex dynamics including chaos [14]. High-frequency dynamical regimes of passive feedback lasers were then reported by Ushakov et al. in 2004 [235] using an integrated distributed feedback device that allows the control of the feedback phase.

7.3.3 Numerical simulations and bridges

Although the single-ECM solution (7.81) exactly satisfies the LK equations, the two-ECM solution of the form

$$Y = A_1 \exp\left(i(\Omega_1 - \Omega_0)t\right) + A_2 \exp\left(i(\Omega_2 - \Omega_0)t\right) \qquad (7.86)$$

is not an exact solution of the LK equations. However, (7.86) can be the leading approximation of an asymptotic solution valid for large T. Specifi-

---

[3]The first workshop on "Dynamics of Semiconductor Lasers" occurred at the WIAS in September 1999. It gathered mathematicians, physicists, and engineers with very different backgrounds. But the expectations were high. H. J. Wünsche asked the mathematicians to tell "how dynamical systems may provide new ideas for 40 GHz lasers". Most mathematicians were scared, having no understanding of what, for example, a DFB laser was. But the interaction between mathematicians and physicists persisted almost every year at the WIAS workshops. Today, the topics discussed go beyond laser problems and deal with the delayed synchronization of oscillators or the stabilization of unstable states by a delayed feedback control.

cally, we seek a perturbation solution of the form

$$Y = Y_0(t) + \varepsilon Y_1(t) + \cdots \tag{7.87}$$
$$Z = Z_0(t) + \varepsilon Z_1(t) + \cdots, \tag{7.88}$$

where $\varepsilon \equiv T^{-1}$. From Eqs. (7.79) and (7.80), the leading-order problem as $\varepsilon = 0$ is given by

$$\frac{dY_0}{dt} = (1 + i\alpha)Z_0 Y_0 + \eta \exp(-i\Omega_0\theta)Y_0(t - \theta), \tag{7.89}$$

$$\frac{dZ_0}{dt} = 0. \tag{7.90}$$

Equation (7.90) implies that

$$Z_0 = C \tag{7.91}$$

is an unknown constant. Equation (7.89) with $Z_0 = C$ is then linear and admits the solution (7.86) only at particular values of the feedback rate ($\eta = \eta_c$). At $\eta = \eta_c$, two single ECMs admit the same value of $Z = B$ and the frequencies

$$\Delta_1 = \Delta_{1c} \quad \text{and} \quad \Delta_2 = \Delta_{2c}. \tag{7.92}$$

The two-ECM solution (7.86) exhibits two amplitudes A_1 and A_2 which are undetermined at this order of the perturbation analysis. In order to determine equations for A_1 and A_2, we need to investigate the higher-order problem for Y_1 and Z_1 and apply solvability conditions [57]. The analysis is long and tedious because of transcendental equations for the unknown amplitudes A_1, A_2 and the first-order correction of the frequencies Ω_1 and Ω_2. But the bifurcation results are relatively simple. The two-ECM solution (7.86) belongs to a closed branch of solutions connecting two Hopf bifurcation points each located on a different single ECM branch. They are the bridges found numerically (see Figure 7.16). The bifurcation diagram of the maxima and minima of $|Y|$ obtained by integrating the LK equations for gradually increasing (or decreasing) values of η is shown in the top figure for the same values of the parameters as in Figure 7.15. The figure shows successive stable ECM branches each undergoing a Hopf bifurcation. The same diagram now obtained by a continuation method is shown in the bottom figure. Only the maxima are shown. The figure reveals bridges connecting distinct Hopf bifurcation points. The single-ECM solution (7.81) admits a constant intensity given by $I = A^2$ but the two-ECM solution (7.86) exhibits time-periodic intensity oscillations of the form

$$I = |A_1|^2 + |A_2|^2 + 2|A_1||A_2|\cos\left((\Delta_{1c} - \Delta_{2c})\theta^{-1}t + \phi\right), \tag{7.93}$$

where ϕ is a constant phase. The period of the oscillations is the mode-beating period

$$P_{MB} = 2\pi\theta |\Delta_{1c} - \Delta_{2c}|^{-1} \tag{7.94}$$

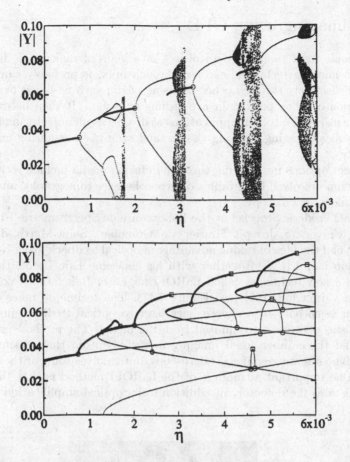

Figure 7.16: Top: Bifurcation diagram of the stable solutions obtained by integration in time. Bottom: Bifurcation diagram of the stable and unstable steady and periodic solutions obtained by a continuation method (reprinted Fig. 1 with permission from Pieroux et al. [190] copyright 2001 American Physical Society).

which is clearly proportional to the delay θ. Numerical bifurcation studies suggest that bridges are either unstable or are partially stable [57, 190, 59]. However, stable bridges are possible if α is sufficiently low ($\alpha \leq 1$) [59]. For an arbitrary value of $\alpha > 1$, a stable bridge may change its stability at a torus bifurcation point as we increase the feedback rate. The torus bifurcation leads to quasiperiodic oscillations with two distinct frequencies. The first and second frequencies are the bridge intensity frequency P_{MB}^{-1} and the RO frequency $\omega_r = \sqrt{2P/T}$ [190, 59].

7.4   Imaging using OFB

Few people need to inspect an object in a glass of milk [134]. If, however, an imaging technique can see through milk, it probably can image objects effectively through other diffusing media such as blood or even a suspension of silica powder in a polishing workshop. It then becomes an effective inspection tool in applications as diverse as manufacturing inspection, medical imaging of living tissues, and even tasks requiring undersea visibility.

Current options for imaging through diffusing media include techniques such as time-resolved holography, optical coherence tomography, and scanning confocal microscopy.

In 1999 Frédéric Stoeckel at the Laboratoire de Spectrométrie Physique of the Université Joseph Fourier de Grenoble (Saint-Martin-d'Héres, France) had the idea of taking advantage of optical feedback using Nd:YAG microchip lasers [134]. Together with his colleague Eric Lacot, they developed a new technique called LaROFI for Laser Relaxation Oscillation Frequency Imaging [135] (see Figure 7.17). The technique relies on the resonant sensitivity of a short-cavity laser to optical feedback produced by ballistic photons retrodiffused from the media. The method produces two- and three-dimensional imaging in turbid media that is similar to heterodyne scanning confocal microscopy, but resolves some of the limitations. One important advantage of the LaROFI method is that the laser source is also the detector. In addition to its optical-amplification duties,

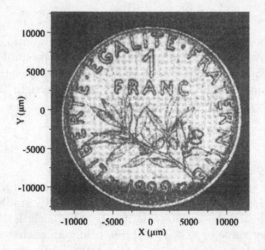

Figure 7.17: Two-dimensional (262 × 262 pixels) image of a French 1-franc piece using the laser relaxation oscillation frequency imaging technique. The pixel dimensions are 100 μm×100 μm (from Lacot et al. [135]).

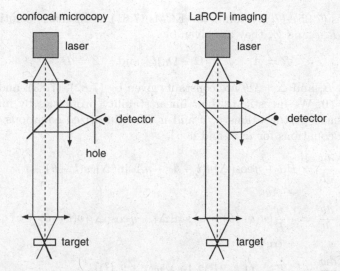

Figure 7.18: In confocal microscopy, spatial filtering is controlled by the size of the hole and the quality of the detection depends on the detector. In LFI and LaROFI imaging, spatial filtering is achieved by selection of one laser mode and the quality of the detection depends on the laser.

it provides self-aligned spatial and temporal coherent detection (acts as both a spatial and temporal filter). See Figure 7.18. Another novelty of the LaROFI technique is that the frequency of the intensity relaxation oscillations is measured together with the intensity of the laser field. This provides a $100 \times$ higher sensitivity compared to previous techniques based on external cavity frequency measurements (LFI for Laser Feedback Interferometry). In this section, we determine approximations of the laser intensity and the laser relaxation oscillation frequency that we compare with the experimental data.

7.4.1 Stability analysis

The LK equations (7.79) and (7.80) describe the response of a single-mode laser subject to optical feedback from a distant mirror. Introducing the amplitude and the phase of the field, $Y = R\exp(i\phi)$, these equations with $\alpha = 0$ (we are dealing with microchip solid state lasers where α is zero) can be rewritten as

$$\frac{dR}{dt} = ZR + \eta R(t-\theta)\cos(\phi(t-\theta) - \phi - \Omega_0\theta), \qquad (7.95)$$

$$\frac{d\phi}{dt} = \alpha Z + \eta\frac{R(t-\theta)}{R}\sin(\phi(t-\theta) - \phi - \Omega_0\theta), \qquad (7.96)$$

$$T\frac{dZ}{dt} = P - Z - (1+2Z)R^2. \qquad (7.97)$$

Equations (7.95)–(7.97) admit the ECMs (7.81) as the basic solutions. In terms of R, ϕ, and Z, they are given by

$$R = A, \quad \phi = (\Omega - \Omega_0)t \quad \text{and} \quad Z = B, \tag{7.98}$$

where A, B, and $\Delta = \Omega\theta$ are constants given by (7.82), (7.83), and (7.84) with $\alpha = 0$. We investigate their linear stability properties by introducing the small perturbations u, v and w. The linearized equations are the following equations for u, v, and w

$$\frac{du}{dt} = Bu + \eta\cos(\Delta)u(t-\theta) + \eta A\sin(\Delta)(v(t-\theta) - v) \tag{7.99}$$
$$+ Aw,$$

$$\frac{dv}{dt} = -\frac{\eta}{R}(u(t-\theta) - u)\sin(\Delta) + \eta\cos(\Delta)(v(t-\theta) - v)$$
$$+ \alpha w, \tag{7.100}$$

$$\frac{dw}{dt} = -T^{-1}\left[(1+2B)2Au + w(1+2A^2)\right]. \tag{7.101}$$

We solve these equations by looking for a solution of the form $u = a\exp(\lambda t)$, $v = b\exp(\lambda t)$ and $w = c\exp(\lambda t)$. We then obtain the following problem for the coefficients a, b, and c

$$\lambda\begin{pmatrix} a \\ b \\ c \end{pmatrix} = L\begin{pmatrix} a \\ b \\ c \end{pmatrix}, \tag{7.102}$$

where the Jacobian matrix L is defined by

$$L \equiv \begin{pmatrix} \eta\cos(\Delta)F & \eta A\sin(\Delta)F & A \\ -\frac{\eta\sin(\Delta)}{A}F & \eta\cos(\Delta)F & \alpha \\ -(1+2B)2A\varepsilon & 0 & -(1+2A^2)\varepsilon \end{pmatrix}, \tag{7.103}$$

where

$$F \equiv \exp(-\lambda\theta) - 1 \quad \text{and} \quad \varepsilon \equiv T^{-1}. \tag{7.104}$$

A nontrivial solution is possible only if λ satisfies the condition $\det(L - \lambda I) = 0$. This condition leads to the characteristic equation for the growth rate λ,

$$0 = \left[-(1+2A^2)\varepsilon - \lambda\right]\left\{\begin{matrix}[\eta\cos(\Delta)F - \lambda]^2 \\ +\eta^2\sin^2(\Delta)F^2\end{matrix}\right\}$$
$$+ (1+2B)2A^2\varepsilon\left[\eta\cos(\Delta)F - \lambda\right]. \tag{7.105}$$

7.4.2 Low feedback rate approximation

Equation (7.105) is hard to solve even numerically. Several approximations have been investigated in the past (see review in [58]). In this section, we propose to investigate the solution of Eq.(7.105) for low values of η. If $\eta\theta$ is small, there is only one ECM. From (7.83), (7.84), and (7.82), we find the simple approximation

$$\Delta = \Omega_0\theta + O(\eta\theta), \quad A^2 = P + O(\eta) \quad \text{and} \quad B = O(\eta). \tag{7.106}$$

If $\eta = 0$, the characteristic equation (7.105) reduces to

$$\lambda\left[\lambda^2 + \lambda(1 + 2P)\varepsilon + 2P\varepsilon\right] = 0 \tag{7.107}$$

which we recognize as the characteristic equation for the solitary laser. For small ε and $\lambda \neq 0$, Eq. (7.107) has the solution

$$\lambda = \pm i\sqrt{2P\varepsilon} - \varepsilon\frac{1 + 2P}{2} + O(\varepsilon^{3/2}). \tag{7.108}$$

The leading term is the RO frequency defined by

$$\omega_r \equiv \sqrt{2P\varepsilon}. \tag{7.109}$$

The expression (7.108) motivates seeking a solution of (7.105) of the form

$$\lambda = \varepsilon^{1/2}\lambda_0 + \varepsilon\lambda_1 + ... \tag{7.110}$$

and in order to balance terms with η in Eq. (7.105), we assume η as an $O(\varepsilon)$ quantity. Specifically, we expand η as

$$\eta = \varepsilon\eta_1 + ... \tag{7.111}$$

Introducing (7.110) and (7.111), taking into account (7.109), we equate to zero the coefficients of each power of $\varepsilon^{1/2}$. The first two problems are

$$O(\varepsilon^{3/2}) : 0 = -\lambda_0^3 - 2P\lambda_0, \tag{7.112}$$

$$O(\varepsilon^2) : 0 = -(3\lambda_0^2 + 2P\varepsilon)\lambda_1 + 2\lambda_0^2\eta_1\cos(\Omega_0\theta)F_0$$
$$-(1 + 2P)\lambda_0^2 + 2PF_0\eta_1\cos(\Omega_0\theta), \tag{7.113}$$

where

$$F_0 \equiv \exp(-\varepsilon^{1/2}\lambda_0\theta) - 1 \tag{7.114}$$

and we have assumed $\varepsilon^{1/2}\theta = O(1)$. From Eq. (7.112) and then Eq. (7.113), we determine λ_0 and λ_1. Together, the growth rate λ is given by

$$\lambda \simeq \pm i\omega_r + \frac{1}{2}\begin{bmatrix} -\varepsilon(1 + 2P) \\ -2\sin^2(\omega_r\theta)\eta\cos(\Omega_0\theta) \\ \mp i\sin(\omega_r\theta)\eta\cos(\Omega_0\theta) \end{bmatrix}. \tag{7.115}$$

Lacot et al. [135] designed the LaROFI imaging technique using the expression (7.115). See Figure 7.19. Specifically, the method determines the

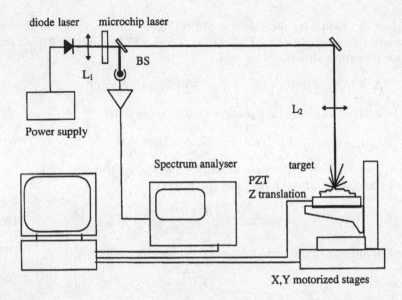

Figure 7.19: The laser beam is focused on a target. Only photons backscattered from points located near the center of the laser beam on the target are reinjected by mode matching into the laser. The laser dynamics is modified by the interference effects taking place between the backscattered field and the standing wave inside the laser cavity. This interference effect depends on the reflectivity, distance, and motion of the target. The laser output power is detected by a photodetector and the laser relaxation frequency is determined by a spectrum analyzer. In order to obtain an image, a micrometric translation unit combined with a PZT moves the target (from Lacot et al. [135]).

modification of the relaxation oscillation frequency of the laser as the feedback rate increases. In the case of constructive interference,

$$\cos(\Omega_0 \theta) = 1, \tag{7.116}$$

the ECM solution 7.106) is stable because $\mathrm{Re}(\lambda) < 0$. The imaginary part provides the correction to the RO frequency ω_r due to optical feedback. This relative change of the RO frequency is given by

$$\frac{\omega_{OF} - \omega_r}{\omega_r} = -\frac{\eta}{2\omega_r}\sin(\omega_r \theta). \tag{7.117}$$

Furthermore, if $\omega_r \theta$ is small, we have $\sin(\omega_r \theta) \simeq \omega_r \theta$, and the expression (7.117) can be further simplified. In terms of the original parameters, it leads to

$$F \equiv \frac{\Omega_{OF} - \Omega_r}{\Omega_r \sqrt{R_{eff}}} = -\frac{\gamma_c}{2}\tau = -\frac{\gamma_c}{c}d. \tag{7.118}$$

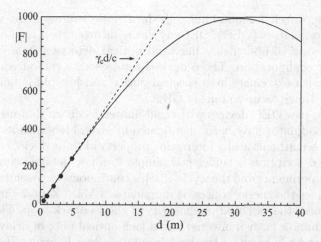

Figure 7.20: Modified RO frequency due to optical feedback. The figure represents the relative change of the relaxation frequency $|F|$ as a function of the distance laser target d.

Ω_r and τ are defined by

$$\Omega_r \equiv \sqrt{\gamma_1 \gamma_c (P-1)} \quad \text{and} \quad \tau \equiv \frac{2d}{c}, \qquad (7.119)$$

where the population inversion damping rate $\gamma_1 = 1/(255 \ \mu s) = 3.9215 \times 10^3 \ s^{-1}$, the cavity damping rate $\gamma_c = 1.55 \times 10^{10} \ s^{-1}$, the pump parameter above threshold $P = 2$, and the effective feedback reflectivity $R_{eff} = 10^{-4}$. d is the distance laser target and $c = 3 \times 10^8 \ ms^{-1}$ is the speed of light. The expression (7.118) indicates that the relative change of the RO frequency is linearly proportional to the distance d for low values of d (broken line in Figure 7.20) Figure 7.20 also shows the frequency correction $|F|$ obtained from (7.117) but without the approximation $\omega_r \theta$ is small (full line). The dots are experimental points. Note that the increase of $|F|$ does not remain linear but exhibits a maximum near $d = 30$ m. This behavior resulting from the sine function in (7.117) remains to be checked experimentally.

7.5   Optoelectronic oscillators

High repetition rate pulse sources are usually implemented by active mode locking of fiber or diode lasers which requires a microwave-driving source whose phase noise determines or limits the resultant jitter.[4] A completely

---

[4] Jitter is an unwanted variation of the signal characteristics of the laser output. Jitter may be seen in the interval between successive pulses, or the amplitude, frequency, or phase of successive cycles.

different approach for obtaining sustained pulse sources is to use optoelectronic oscillators (OEO). OEOs typically incorporate a nonlinear modulator, an optical-fiber delay line, and optical detection in a close-loop
resonating configuration. These devices can generate radio-frequency oscillations with extremely high spectral purity and low phase noise in the
microwave range as up to tens of GHz.

Recently, new OEO devices with band-limited feedback that use a Mach-
Zehnder modulator have been investigated in several laboratories [12, 109,
128, 129]. A mathematically interesting property of these devices is the fact
that their description is sufficiently simple to allow detailed comparisons
between experiments and theory. In [128], a continuous-wave semiconductor
laser provides the energy source. It illuminates a Mach-Zehnder modulator
that produces the essential nonlinearity of the feedback loop. The output
of the modulator is then injected into a long optical fiber of delay time τ_D
and a photodiode converts the light into an electrical current. The photodiode admits a low and a high cutoff frequencies f_L and f_H, respectively.
Finally, a radio-frequency amplifier converts the signal from the photodiode
into an electrical voltage that is fed back in the Mach-Zehnder modulator.
In dimensionless form, the voltage of the feedback loop satisfies the following integro-delay differential equation [128]

$$x + \tau \frac{dx}{dt} + \theta^{-1} \int_0^t x(s)ds = \beta \left[\cos^2 \left(x(t - \tau_D) + \phi \right) - \cos^2 (\phi) \right], \quad (7.120)$$

where the feedback amplitude β and the feedback phase ϕ are two independent control parameters. The time constants θ and τ are directly related
to the cutoff frequencies f_L and f_H, respectively. Eq. (7.120) differs from
Ikeda equation (1.26) by the presence of the integral term. In [128], the
OEO exhibits different orders of magnitude for the three time constants,
namely,

$$\tau = 25 \text{ ps}, \quad \tau_D = 30 \text{ ns}, \quad \text{and} \quad \theta = 5 \text{ } \mu\text{s},$$

suggesting the possibility of multiple time-scale regimes. Figure 7.21 shows
two different experimental outputs for the power, proportional to $\beta \cos^2$
$(x + \phi)$, as a function of time, measured in μs. The square-wave oscillations
(bottom) are reminiscent of Ikeda square-wave oscillations but the low-
frequency asymmetric regime (top) is new.

To have an analytical insight of this particular solution, we reformulate
Eq. (7.120) as a system of two coupled first order DDEs. To this end, we
differentiate Eq. (7.120) once, introduce $x = dz/dt$, and obtain

$$x' = z, \quad (7.121)$$
$$\tau z' = -z - \theta^{-1}x - \beta \sin \left(2x(t - \tau_D) + 2\phi \right), \quad (7.122)$$

where prime means differentiation with respect to t. We next anticipate
that the solution exhibits a basic time scale proportional to θ and introduce

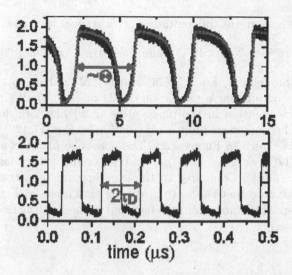

Figure 7.21: Two distinct oscillatory outputs are observed for $\beta = 1.3$. Top: $\sin(2\phi) < 0$ and the oscillations are asymmetric and exhibit a period proportional to θ. The thin grey line corresponds to the numerical solution of the integro-differential equation for $\beta = 1.3$, $\phi = -0.57$, and using $F = 1.632\beta \cos^2(x+\phi)$. Bottom: $\sin(2\phi) > 0$ and the oscillations are nearly square-wave with a period close to $2\tau_D$. Experiments courtesy of Laurent Larger.

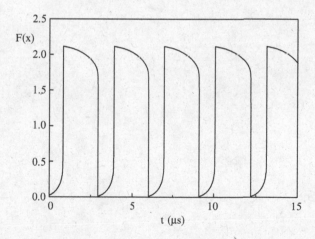

Figure 7.22: Limit-cycle solution of the coupled first order equatuions with $\delta = 0$.

$s \equiv t/\theta$ and $v \equiv \theta z$. From Eqs. (7.121) and (7.122), we find

$$x' = v, \tag{7.123}$$

$$\varepsilon v' = -v - x - \beta \sin\left(2x(s-\delta) + 2\phi\right), \tag{7.124}$$

where prime now means differentiation with respect to s and

$$\varepsilon \equiv \tau\theta^{-1} = 2.46 \times 10^{-6} \quad \text{and} \quad \delta \equiv \tau_D\theta^{-1} = 8.43 \times 10^{-3}.$$

Neglecting the delay δ, Eqs. (7.123) and (7.124) reduce to two ordinary differential equations for a relaxation oscillator ($\varepsilon \ll 1$). We may analyze the limit-cycle solution in the phase-plane (x, v) like Van der Pol limit-cycle [22]. The numerical limit-cycle solution of Eqs. (7.123) and (7.124) with $\delta = 0$ is shown in Figure 7.22. The parabolic plateaus are functions of s ($\Delta s = O(1)$ or $\Delta t \sim \theta \sim \mu s$) and are connected by fast transition layers ($\Delta s = O(\varepsilon)$). Compared to the experimental time trace in Figure 7.21 (top), we note that the analytical transition layers in Figure 7.22 are much sharper. This comes from the fact that we neglect the delay δ, in first approximation.

8

Phase equations

Systems of coupled oscillators are a suitable approach for studying rhythmic behavior at many levels of biological organization, from populations of molecules to populations of organisms. Because their collective response is rich and varied, there exists a large literature on the synchronization properties of coupled oscillators. Communication between different units of a population may take time (as electrical signals propagating between neurons) and scientists are now investigating the effect of a delayed coupling. This chapter does not fall into the trap of reviewing this emerging field of research. But some salient features of the response of delayed coupled oscillators can be understood by analyzing just two coupled oscillators. The simplest approach that describes the synchronization of two oscillators is based on single Delay Differential Phase Equations (DDPEs). Although simple, they offer challenging new bifurcation problems because both bounded and unbounded solutions are acceptable.

In Chapter 5, we discussed the case of a chemical oscillator subject to weak delayed feedback. It is not an example of two coupled oscillators but it leads to a DDPE that is worthwhile to discuss. Close to a supercritical Hopf bifurcation, the effect of a weak delayed feedback on a limit-cycle oscillator can be described by the following DDPE,

$$\phi' = \omega + \mu \sin\left[\phi - \phi(t - \tau)\right], \qquad (8.1)$$

where ϕ represents the phase of the limit-cycle oscillator. This equation is one form of Adler's equation with delay.[1] Another form is introduced below.

---

[1] Robert Adler (1913–2007) is best known as the co-inventor of the television remote control using ultrasonic waves. But in the 1940s, he and others at Zenith Corporation

The first term in the right-hand side of Eq. (8.1) represents the nonlinear correction of the frequency of the free oscillator. The second term in Eq. (8.1) is the contribution of the delayed feedback. Substituting $\phi = \Omega t$ into Eq. (8.1) leads to the following equation for Ω,

$$\Omega = \omega + \mu \sin(\Omega \tau) \tag{8.2}$$

from which we extract the implicit solution

$$\tau = \frac{1}{\Omega} \arcsin(\frac{\Omega - \omega}{\mu}). \tag{8.3}$$

Provided μ is sufficiently large, the period $T \equiv 2\pi/\Omega$ as a function of τ exhibits hysteresis as shown in Figure 8.1. Using the conditions for a nascent hysteresis $(d\tau/d\Omega = d^2\tau/d\Omega^2 = 0)$, we find

$$1 = \mu\tau\cos(\Omega\tau) \quad \text{and} \quad \sin(\Omega\tau) = 0. \tag{8.4}$$

Together with Eq. (8.2), this determines the critical value of $\mu = \mu_c \equiv \omega/2\pi$ above which a bistable response diagram is possible. The solutions $\phi = \Omega t$ are called the locked states because the period $T = 2\pi\Omega^{-1}$ of the free oscillations try to match the delay τ.

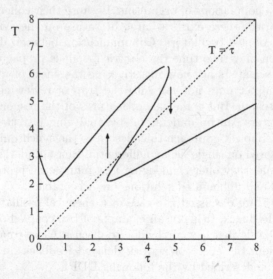

Figure 8.1: The values of the parameters are $\omega = 2$ and $\mu = 1$.

were interested in reducing the number of vacuum tubes in an FM radio. The possibility that a locked oscillator might offer a solution inspired his 1946 paper "A Study of Locking Phenomena in Oscillators." Adler's work concerned a single nonlinear phase oscillator. Later the idea was exploited and generalized to describe a number of similar coupled oscillators. Some authors refer to "Adler-type equations" in connection with models of the form (8.1) where the sine function is the nonlinear function.

8.1   Weakly coupled oscillators

For two weakly coupled oscillators, the predominant effect is a synchronization of the frequencies of the individual oscillators to a single common frequency once the coupling strength exceeds a certain threshold, while the amplitudes remain unaffected. Schuster and Wagner [212] and Niebur et al. [177] first investigated the response of two delayed coupled oscillators using the following equations

$$\phi_1' = -\frac{\Delta}{2} - \kappa \sin(\phi_1 - \phi_2(t - \tau)), \tag{8.5}$$

$$\phi_2' = \frac{\Delta}{2} - \kappa \sin(\phi_2 - \phi_1(t - \tau)). \tag{8.6}$$

We have restricted the model equations to the case of symmetric detuning where $\omega_2 = -\omega_1 = \Delta/2$. $\Delta > 0$ is considered as the bifurcation parameter. The choice of a symmetric detuning is motivated by the analysis of a laser experiment considered in the next section.

8.1.1   Basic solutions

Equations (8.5) and (8.6) admit a constant phase solution of the form

$$\phi_1 = 0 \quad \text{and} \quad \phi_2 = \alpha. \tag{8.7}$$

From Eqs. (8.5) and (8.6), we find α as

$$\alpha = \arcsin(\frac{\Delta}{2\kappa}) \quad (\Delta \leq 2\kappa). \tag{8.8}$$

We next investigate its stability. From the linearized equations, we determine the characteristic equation for the growth rate λ

$$\lambda^2 + 2\kappa\lambda \cos(\alpha) + \kappa^2 \cos^2(\alpha)(1 - \exp(-2\lambda\tau)) = 0. \tag{8.9}$$

The case of purely real eigenvalues can be analyzed by solving the quadratic equation in $\cos(\alpha)$. We find two possible solutions given by

$$\cos(\alpha) = -\frac{\lambda}{\kappa}(1 + \exp(-\lambda\tau))^{-1}, \tag{8.10}$$

$$\cos(\alpha) = -\frac{\lambda}{\kappa}(1 - \exp(-\lambda\tau))^{-1}. \tag{8.11}$$

The two branches of solutions are shown in Figure 8.2 (left). We note from the figure that $\lambda\tau < 0$ if $\cos(\alpha) > 0$. This condition is thus a necessary condition for stability. The open circle marks a simple zero eigenvalue that corresponds to a limit point of the $\alpha = \alpha(\Delta\tau)$ branch (see Figure 8.2 (right); the limit point is located at $(\Delta\tau, \alpha) = (2\kappa\tau, \pi/2)$. The filled circle

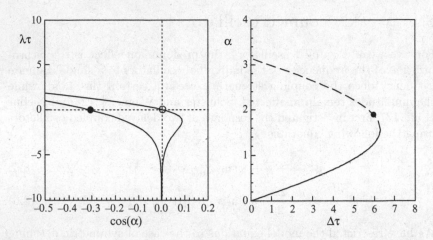

Figure 8.2: Stability of the constant phase solution. The values of the parameters are $\kappa = \pi$ and $\tau = 1$. Left: Real eigenvalues $\lambda\tau$ as a function of $\cos(\alpha)$. The full and open circles indicate simple zero eigenvalues. Right: Bifurcation diagram of the constant phase solution $\phi_1 = 0$, $\phi_2 = \alpha$. Based on the sign of the real eigenvalues, the lower and upper branches are stable and unstable, respectively. The dot marks a bifurcation point to the frequency locked solution $\phi_1 = \Omega t$ and $\phi_2 = \Omega t + \alpha$.

in Figure 8.2 (left) denotes another zero eigenvalue located at $\cos(\alpha) = -1/\kappa\tau$. With $\sin(\alpha)$ obtained from (8.8), we find that this particular point is located at $\Delta = \Delta_c$ where

$$\Delta_c \equiv 2\kappa\sqrt{1 - \frac{1}{\kappa^2\tau^2}} \quad (1 - \frac{1}{\kappa^2\tau^2} > 0) \tag{8.12}$$

(full circle in Figure 8.2 (right)). This point is a bifurcation point to a frequency locked solution of the form

$$\phi_1 = \Omega t \quad \text{and} \quad \phi_2 = \Omega t + \alpha. \tag{8.13}$$

Schuster and Wagner [212] analyzed these solutions. Substituting (8.13) into Eqs. (8.5) and (8.6), we obtain two equations for Ω and α given by

$$\Omega = -\frac{\Delta}{2} - \kappa\sin(\Omega\tau - \alpha), \tag{8.14}$$

$$\Omega = \frac{\Delta}{2} - \kappa\sin(\Omega\tau + \alpha). \tag{8.15}$$

Adding and substracting Eqs. (8.14) and (8.15) lead to expressions for $\cos(\alpha)$ and $\sin(\alpha)$, respectively. We then eliminate α using a trigonometric identity and obtain the solution in the implicit form $\Delta = \Delta(\Omega)$:

$$\Delta = 2\kappa|\cos(\Omega\tau)|\sqrt{1 - \frac{\Omega^2}{\kappa^2\sin^2(\Omega\tau)}}, \tag{8.16}$$

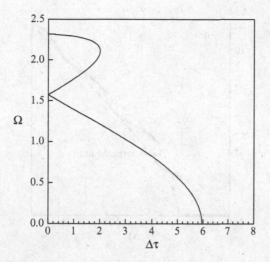

Figure 8.3: Frequency locked solution: $\kappa = \pi$ and $\tau = 1$.

where the expression in the squareroot must be positive. Note that as $\Omega \to 0,\; \Delta \to \Delta_c$ where Δ_c is defined by 8.12. The branch of frequency locked solutions is shown in Figure 8.3. Its stability needs to be analyzed numerically. Simulations of Eqs. (8.5) and (8.6), however, suggest that the constant phase solution (8.7) is the only stable solution for $0 < \Delta\tau < \kappa\tau$.

Motivated by laser experiments, Wünsche et al. [245] became interested in other forms of synchronization. Adding Eqs. (8.5) and (8.6), we obtain

$$\phi_1' + \phi_2' = -2\kappa \sin\left(\frac{\phi_1 - \phi_2(t-\tau) + \phi_2 - \phi_1(t-\tau)}{2}\right)$$
$$\times \cos\left(\frac{\phi_1 - \phi_2(t-\tau) - \phi_2 + \phi_1(t-\tau)}{2}\right). \qquad (8.17)$$

Equation (8.17) admits the particular solution

$$\phi_2 + \phi_1 = C, \qquad (8.18)$$

where C is a constant (the left-hand side and the sine function in the right-hand side are zero). With (8.18), the two equations for ϕ_1 and ϕ_2 reduce to a single equation. Introducing $\phi_1 = C/2 - \Phi$ and $\phi_2 = C/2 + \Phi$, Eq. (8.6) can be rewritten as

$$\Phi' = \frac{\Delta}{2} - \kappa \sin(\Phi + \Phi(t-\tau)). \qquad (8.19)$$

The steady-state solution satisfies $\Phi = \frac{1}{2}\arcsin(\Delta/2\kappa)$ and matches (8.8) with $C = 2\Phi$ and $\alpha = 2\Phi$. From (8.19), we determine the linearized equation and obtain the characteristic equation (8.10), which told us that a stable solution is possible in the interval $0 < \Delta\tau < 2\kappa\tau$.

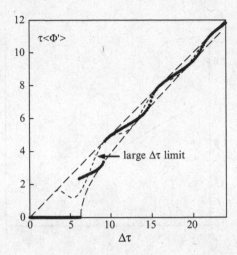

Figure 8.4: Numerical bifurcation diagram: $\kappa\tau = \pi$. The dots are the stable solutions of the phase equation. They have been obtained by progressively increasing $\Delta\tau$ ($0 < \Delta\tau < 24$; $\Phi = 10$ for $-\tau < t < 0$) and by progressively descreasing $\Delta\tau$ ($0 < \Delta\tau < 24$; $\Phi = 0$ for $-\tau < s < 0$). The staircase bifurcation diagram is bounded above by the straight line $\tau < \Phi' >= \Delta\tau/2$ and below by the parabola $\tau < \Phi' >= \sqrt{\Delta^2\tau^2/4 - \kappa^2\tau^2}$.

The numerical bifurcation diagram for the average of $\phi_2' = \Phi'$ is shown in Figure 8.4.[2] As the product $\Delta\tau$ is progressively increased from zero, the staircase bifurcation diagram changes into a snake-like diagram that we would like to capture analytically. To this end, we consider the large $\Delta\tau$ limit of Eq. (8.19). Introducing the new time

$$s \equiv \Delta t, \tag{8.20}$$

Eq. (8.19) becomes

$$\Phi' = \frac{1}{2} - \frac{\kappa}{\Delta}\sin(\Phi + \Phi(s - \Delta\tau)) \tag{8.21}$$

where prime now means differentiation with respect to s. If $\Delta \to \infty$, the leading equation is $\Phi' = 1/2$ implying the solution $\Phi = s/2 + \Theta$ where Θ is a constant. This motivates seeking a solution of the form

$$\Phi = \Omega\frac{s}{2} + \Theta + \Delta^{-1}\Phi_1(s) + \Delta^{-2}\Phi_2(s) + ..., \tag{8.22}$$

where Ω is expanded as

$$\Omega = 1 + \Delta^{-2}\Omega_2 + ... \tag{8.23}$$

---

[2]The average of Φ' is defined as $< \Phi' >= \lim_{t\to\infty} t^{-1}\int^t \Phi'(s)ds = \lim_{t\to\infty} t^{-1}\Phi(t)$.

and takes into account possible corrections to $\Omega = 1$. We have anticipated that the first nonzero correction is $O(\Delta^{-2})$. Ω_2 will be determined by requiring that the functions Φ_1, Φ_2 are bounded periodic 2π-functions of s. Substituting (8.22) and (8.23) into Eq. (8.21), we obtain the following problems for Φ_1 and Φ_2,

$$\Phi_1' = -\kappa \sin(s - \frac{\Delta\tau}{2} + 2\Theta) \tag{8.24}$$

$$\Phi_2' = -\kappa \cos(s - \frac{\Delta\tau}{2} + 2\Theta)(\Phi_1 + \Phi_1(s - \Delta\tau)) - \frac{\Omega_2}{2}. \tag{8.25}$$

The solution of Eq. (8.24) is purely periodic and has the form

$$\Phi_1 = \kappa \cos(s - \frac{\Delta\tau}{2} + 2\Theta) + \Theta_1, \tag{8.26}$$

where Θ_1 is a new constant. The solution of Eq. (8.25), however, will contain a term proportional to s unless we set the average of its right-hand side equal to zero. This is the solvability condition that allows Φ_2 to be bonded and 2π-periodic in s. The condition leads to an expression for Ω_2 given by

$$\Omega_2 = -\kappa^2(1 + \cos(\Delta\tau)). \tag{8.27}$$

In summary, we have found that the average $< \Phi' > = \Omega\Delta/2$ is given by

$$< \Phi' > \simeq \frac{\Delta}{2} - \frac{\kappa^2}{2\Delta}(1 + \cos(\Delta\tau)), \tag{8.28}$$

where the second term explains the snakelike structure of the bifurcation diagram.

8.1.2 Experiments

Two delay-coupled semiconductor lasers have been studied in the regime where the coupling delay is comparable to the time-scales of the internal laser oscillations [245]. Detuning the optical frequencies between the two lasers leads to two distinct responses. See Figure 8.5. For small detuning (-4.1 GHz $< \Delta f < 7$ GHz), the lasers lock onto a common optical frequency and the output intensities are steady and stable. Outside this locking region, the output intensity of both lasers oscillates periodically. The observed oscillation frequencies appear in a discontinuous way. For sufficiently large detunings, the frequency plateaus approach the straight lines $f = \pm |\Delta|$. Numerical simulations of the laser equations allow to give a more detailed description of the staircase distribution of the frequencies (see Figure 8.6).

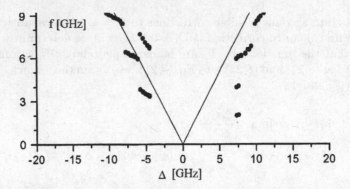

Figure 8.5: Intensity oscillation frequencies f obtained from power spectra as a function of the optical frequency detuning Δ (reprinted Fig. 2b with permission from Wünsche et al. [245], copyright 2005 American Physical Society).

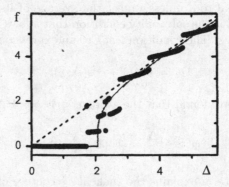

Figure 8.6: Main intensity frequencies obtained numerically from the laser model equations (reprinted Fig. 5b with permission from Wunsche et al. [245] copyright 2005 American Physical Society). All frequencies are in units of $1/2\tau$ (i.e., the roundrip laser 1–laser 2–laser 1 frequency). The frequencies exhibit a typical staircase bifurcation diagram. For sufficiently large detuning Δ, the frequencies f are close to the straight line $f = \Delta$.

8.2   Strongly coupled oscillators

For stronger couplings the amplitudes also play an important role and give rise to interesting behaviors such as partial synchronization or large amplitude oscillations. In this section, we concentrate on the so-called "amplitude death" phenomenon. This designation is somewhat unfortunate inasmuch as it corresponds to a stabilization phenomenon where the delay plays a crucial role. We examine this stability problem by introducing the following model equations,

$$Z_1' = \left[1 + i\omega_1 - |Z_1|^2\right] Z_1 + \kappa\left[Z_2(t - \tau) - Z_1\right], \qquad (8.29)$$

$$Z_2' = \left[1 + i\omega_2 - |Z_2|^2\right] Z_2 + \kappa\left[Z_1(t - \tau) - Z_2\right], \qquad (8.30)$$

where τ is the delay, κ is the coupling strength, ω_j are the intrinsic frequencies of the two oscillators, and $Z_{1,2}$ are complex. The time delay parameter is introduced in the argument of the coupling oscillator (e.g., Z_2 in (8.29)) to physically account for the fact that its phase and amplitude information is received by oscillator Z_1 only after a finite time τ (due to finite propagation speed effects). In the absence of coupling ($\kappa = 0$) each oscillator has a stable limit cycle at $|Z_i| = 1$ on which it moves at its natural frequency ω_j. The coupled equations represent the interaction between two weakly nonlinear oscillators (that are near a Hopf bifurcation) and whose coupling strength is comparable to the attraction of the limit cycles. It is important then to retain both the phase and amplitude response of the oscillators. The state $Z_i = 0$ is a steady-state solution of Eqs. (8.29) and (8.30). For $\kappa = 0$, this equilibrium solution is linearly unstable because the individual oscillators tend to a stable limit-cycle $|Z_i| = 1$. The question is whether a nonzero κ may stabilize the zero solution. The conditions for this phenomenon (called amplitude death) have been studied by Aronson et al. [5] who investigated Eqs. (8.29) and (8.30) with $\tau = 0$. They found the following stability conditions

$$\kappa > 1 \quad \text{and} \quad \Delta = |\omega_2 - \omega_1| > 2\sqrt{2\kappa - 1}. \qquad (8.31)$$

Conditions (8.31) indicate that amplitude death occurs only for sufficiently large values of the coupling strength κ and detuning Δ. In particular, the phenomenon is not possible if the two oscillators are identical ($\Delta = 0$). The question now is whether stabilization of the zero solution for two identical oscillators is possible with a delayed coupling. This problem has been investigated in great detail by Ramana Reddy et al. [196, 197] and we summarize the main results. The linearized equations with $\omega_1 = \omega_2 = \omega$ lead to the following characteristic equation for the growth rate λ,

$$(1 - \kappa + i\omega - \lambda)^2 - \kappa^2 \exp(-2\lambda\tau) = 0, \qquad (8.32)$$

which implies the following two possibilities

$$(1) : 1 - \kappa + i\omega - \lambda + \kappa \exp(-\lambda\tau) = 0, \qquad (8.33)$$

$$(2) : 1 - \kappa + i\omega - \lambda - \kappa \exp(-\lambda\tau) = 0. \qquad (8.34)$$

We now assume that the stability boundaries are determined by a Hopf bifurcation and introduce $\lambda = i\sigma$ into Eqs. (8.33) and (8.34). From the real and imaginary parts, we find

$$(1) : 1 - \kappa + \kappa\cos(\sigma\tau) = 0 \quad \text{and} \quad \omega - \sigma - \kappa\sin(\sigma\tau) = 0, \quad (8.35)$$

$$(2) : 1 - \kappa - \kappa\cos(\sigma\tau) = 0 \quad \text{and} \quad \omega - \sigma + \kappa\sin(\sigma\tau) = 0 \quad (8.36)$$

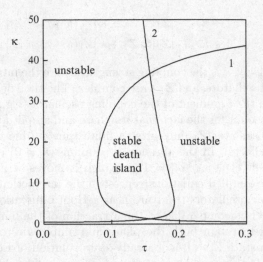

Figure 8.7: Death island. In the closed region, the zero solution is stable whereas it is unstable in all other regions above the lower line $\kappa(\tau) \leq 1$. The value of $\omega = 10$.

which lead to the solutions (in parametric form)

$$(1) : \kappa = 1/(2\sin^2(x/2)) \quad \text{and} \quad \tau = x/(\omega - \cot(x/2)), \quad (8.37)$$
$$(2) : \kappa = 1/(2\cos^2(x/2)) \quad \text{and} \quad \tau = x/(\omega + \tan(x/2)). \quad (8.38)$$

The curves $\kappa = \kappa(\tau)$ are shown in Figure 8.7 and have been obtained by progressively changing x from zero in (8.37) and (8.38). The closed region called "death island" corresponds to a stable zero solution. All other regions above the lower branches of the Hopf curves ($\kappa \leq 1$) correspond to an unstable zero solution. The death region exists only if $\omega > \omega_c \simeq 4.812$. At $\omega = \omega_c$, the death region has shrunk into a point located at $(\tau_c, \kappa_c) \simeq (0.33, 3.06)$. If $\omega < \omega_c$, the two Hopf curves move apart leaving a central unbounded region where the zero solution is stable. See Figure 8.8.

As we progressively increase τ from zero and cross the death island in Figure 8.7, we pass through two Hopf bifurcations. The first bifurcation (line 1) leads to in-phase oscillations and the second bifurcation (line 2) leads to antiphase oscillations. This can be demonstrated by determining the eigenvectors associated with each bifurcation. At the Hopf bifurcation, the small deviations from zero are time-periodic with complex amplitudes $u_1 = u_2$ for the first case (line 1) and with $u_1 = -u_2$ in the second case (line 2).

Is the amplitude death with delay observed in real systems? A number of experiments have addressed this question and carried out laboratory experiments. The simplest example of amplitude death induced by delay was described in 2000 by Ramana Reddy et al. [198] who examined the

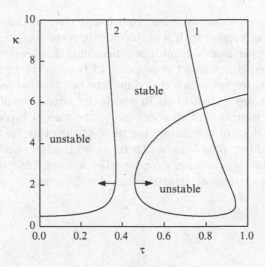

Figure 8.8: $\omega = 4$.

response of two coupled nonlinear electronic circuits. There is good agreement between experiments and theory including the observation of a second death island. Observations of amplitude death were reported by Herrero et al. [92] for a pair of optothermal oscillators but the results were not conclusive on the role of the delay. Takamatsu et al. [224] studied the effects of time-delay in a living coupled oscillator system by looking at the time variation of the thickness of the plasmodium of the slime mold *Physarum polycephalum*. Although amplitude death was not observed, typical delay-induced synchronization patterns such as in-phase and out-of-phase regimes were observed. Tang et al. [226] studied the amplitude death phenomenon (called "death by delay" by the authors) for two semiconductor lasers with optoelectronic feedback mutually coupled optoelectronically. Both the individual feedbacks and the coupling are delayed. The feedback delay induces the limit-cycle oscillations for each laser and the delayed mutual optoelectronic coupling provides the necessary negative feedback for the amplitude death phenomenon. Depending on the delay of the coupling, several regions of amplitude death have been found. On the theoretical side, Atay [6] recently showed that the parameter space of amplitude death for the two-oscillators case is enhanced when the oscillators are connected with time delays distributed over an interval rather than concentrated on a point.

The two coupled oscillator model remains an active topic of research full of surprises. Motivated by a specific two coupled laser problem, Carr et al. [44] discovered that two identical nearly conservative oscillators may exhibit a delay-induced resonance phenomenon where the amplitude of the oscillations become suddenly large. Investigating the possible coupling mechanisms of neural populations, Dahlem et al. [49] considered the delayed coupling of two excitable systems[3] and found that sustained oscillations in antiphase may coexist with a stable steady state (called echo-waves in [231]). On the experimental front, the synchronization properties of chaotic lasers are studied comparing the response of two and three coupled lasers [86, 64]. When three lasers in a row shine into one another in just the right way, they can forge a connection in which the intensities of the first and last lasers vary in unison. This contrasts to the case of two coupled lasers where the variations of one simply lag those of the other by the amount of time it takes light to pass between them, as we might expect. Researchers believe that these synchronization properties are not limited to lasers but applies to other systems of delayed coupled oscillators.

---

[3]In an excitable system, a small perturbation from a stable steady-state may lead to a large pulse of excitation before coming back to rest.

References

[1] G.B. Airy, On the Regulator of the Clock-Work for Effecting Uniform Movement of Equatoreals, Memoirs of the Royal Astronomical Society, vol. 11, 249-267 (1840)

[2] J.M. Albornoz and A. Parravano, Modeling a simple enzyme reaction with delay and discretization, submitted (2008)

[3] U. an der Heiden and M.C. Mackey, The dynamics of production and destruction: Analytic insight into complex behavior, J. Math. Biol. **16**, 75-101 (1982)

[4] A.A. Andronov and S.E. Khaikin Theory of Oscillations, translated and adapted by S. Lefschetz, Princenton University Press, Princeton, NJ (1949)

[5] D. G. Aronson, G. B. Ermentrout, and N. Koppel, Amplitude response of coupled oscillators, Physica D **41**, 403-449 (1990)

[6] F.M. Atay, Distributed delays facilitate amplitude death of coupled oscillators, Phys. Rev. Lett. **91**, 094101 (2003)

[7] M. Atig, J.-P. Dalmont, and J. Gilbert, Saturation mechanism in clarinet-like instruments, the effect of the localised non-linear losses, Applied Acoustics **65**, 1133-1154 (2004)

[8] D. Aubin and A. Dahan Dalmedico, Writing the history of dynamical systems and chaos: longue durée and revolution, disciplines and cultures, Historia Mathematica **29**, 273-339 (2002)

181

[9] Y. Aurégan and C. Depollier, Snoring: Linear stability analysis and in-vitro experiments, J. of Sound and Vibration **188**, 39–54 (1995)

[10] V. Averina, I. Kolmanovsky, A. Gibson, G. Song, and E. Bueler, Analysis and control of delay-dependent behavior of engine air-to-fuel ratio, Proceedings of 2005 IEEE Conference on Control Applications (CCA 2005) 28, 1222–1227 (2005)

[11] S. Bauer, O. Brox, J. Kreissl, G. Sahin, and B. Sartorius, Optical microwave source, Electron. Lett. **38**, 334-335 (2002)

[12] J.N. Blakely, L. Illing, and D.J. Gauthier, High speed chaos in an optical feedback system with flexible timescales, IEEE J. Quantum Electron. **40**, 299 (2004)

[13] O. Brox, S. Bauer, M. Radziunas, M. Wolfrum, J. Sieber, J. Kreissl, B. Sarorius, and H.-J. Wünsche, High-frequency pulsations in DFB lasers with amplified feedback, IEEE J. Quant. Electron. **39**, 1381-1387 (2003)

[14] S. Bauer, O. Brox, J. Kreissl, B. Sartorius, M. Radziunas, J. Sieber, H.-J. Wünsche, and F. Henneberger, Phys. Rev. E **69**, 016206 (2004)

[15] R.B. Banks, Towing Icebergs, Falling Dominoes, and Other Adventures in Applied Mathematics, Princeton University Press, Princeton NJ (1998)

[16] C.T.H. Baker, G.A. Bocharov, C.A.H. Paul, and F.A. Rihan, Modelling and analysis of time-lags in cell proliferation, Numerical analysis report **313** (1997) available at: http://www.ma.man.ac.uk/MCCM/MCCM.html

[17] R.L. Bar-Or, R. Maya, L.A. Segel, U. Alon, A.J. Levine, and M. Oren, Generation of oscillations by the p53-Mdm2 feedback look: A theoretical and experimental study, Proc. Nat. Acad. Sci. USA **97**, 11250-11255 (2000)

[18] J.J. Batzel and H.T. Tran, Modeling instability in the control system for human respiration: applications to infant non-REM sleep, Appl. Math. and Comp. **110**, 1-51 (2000)

[19] J. Bélair and M.C. Mackey, Consumer memory and price fluctuations in commodity markets: An integrodifferential model, J. of Dyn. and Diff. Eq. **1**, 299-325 (1989)

[20] R.E. Bellman and J.M. Danskin, A Survey of the Mathematical Theory of Time Lag, Retarded Control, and Hereditary Processes, Rand Corporation (1954)

[21] R.E. Bellman and K.L. Cooke, Differential-Difference Equations, Academic Press, New York (1963)

[22] C.M. Bender and S.A. Orszag, Advanced Mathematical Methods for Scientists and Engineers, McGraw-Hill, New York (1978)

[23] S. Bernard, B. Cajavec, L. Pujo-Menjouet, M.C. Mackey, and H. Herzel, Modeling transcriptional feedback loops: The role of Gro/TLE1 in Hes1 oscillations, Phil. Trans. Royal Soc. **364**:1155-1170 (2006)

[24] C. Beta, M. Bertram, A.S. Mikhailov, H.H. Rotermund, and G. Ertl, Controlling turbulence in a surface chemical reaction by time-delay autosynchronization, Phys. Rev. E **67**, 046224 (2003)

[25] A. Beuter, L. Glass, M. Mackey, and M. Titcombe, Eds., Nonlinear Dynamics in Physiology and Medicine, Springer-Verlag, New York (2003)

[26] P. Berg, A. Mason, and A. Woods, Continuum approach to car-following models, Phys. Rev. E **61**, 1056 (2000)

[27] S. Bielawski, D. Derozier, and P. Glorieux, Experimental characterization of unstable periodic orbits by controlling chaos, Phys. Rev. A **47**, R2492-R2495 (1993)

[28] S. Bielawski, D. Derozier, and P. Glorieux, Controlling unstable periodic orbits by a delayed continuous feedback, Phys. Rev. E **49**, R971-R974 (1994)

[29] O. Bilous and N. Admundson, Chemical reactor stability and sensitivity, AI ChE Journal **1**, 513-521 (1955)

[30] C. Bissel and A.A. Andronov and the development of Soviet control engineering, IEEE Control Systems Magazine **18**, 56-62 (1998)

[31] E. Boe and H.-C. Chang, Dynamics of delayed systems under feedback control, Chem. Eng. Science **44**, 1281–1294 (1989)

[32] R. Boucekkine, D. de la Croix, and O. Licandro, Modelling vintage structures with DDEs: Principles and applications, Math. Popul. Stud. **11**, 151-179 (2004)

[33] M. Brackstone and M. McDonald, Car-following: A historical review, Transport. Res. Part F **2**, 181-196 (1999)

[34] N.F. Britton, Essential Mathematical Biolgy, Springer Undergraduate Mathematics Series, Springer-Verlag, London 2003

184   References

[35] A.J. Brown, Influence of oxygen and concentration on alcoholic fermentation, J. Chem. Soc. Trans. **61,** 369-385 (1892)

[36] A.J. Brown, Enzyme action, J. Chem. Soc. (Trans.) **81,** 373-388 (1902)

[37] A.H. Brown,Circumnutations: From Darwin to space flight, Plant Physiology **101,** 345-348 (1993)

[38] P. Brunovský, A. Erdélyi, and H.-O. Walther, On a model of a currency exchange rate - Local stability and periodic solutions, J.Dyn.Diff.Eq. **16,** 393-432 (2004)

[39] J.L. Cabrera and J.G. Milton, On-Off intermittency in a human balancing task, Phys. Rev. Lett. **89,** 158702 (2002)

[40] S.A. Campbell and J. Bélair, Resonant codimension two bifurcation in the harmonic oscillator with delayed forcing, Can. Appl. Math. Quart. **7,** 217-238 (1999)

[41] S.A. Campbell and V.G. LeBlanc, Resonant Hopf-Hopf interactions in delay differential equations, J. Dyn. and Diff. Eq. **10,** 327-346 (1998)

[42] S.A. Campbell, Resonant codimension two bifurcation in a neutral functional differential equation, Nonl. Analysis, Theory, Meth. Appl. **30,** 4577-4584 (1997)

[43] S.A. Campbell, J. Bélair, T. Ohira, and J. Milton, Complex dynamics and multistability in a damped harmonic oscillator with delayed negative feedback, Chaos **5,** 1-6 (1995)

[44] T.W. Carr, I.B. Schwartz, M.Y. Kim, and R. Roy, Delayed-mutual coupling dynamics of lasers: scaling laws and resonances, SIAM J. Dynamical Systems **5,** 699-725 (2006); T.W. Carr, Coupling-induced resonance in two mutually and asymmetrically coupled oscillators, Phys. Rev. E submitted (2008)

[45] T.J. Case, An Ilustrated Guide to Theoretical Ecology, Oxford University Press, Oxford (2000)

[46] C. Clanet, On large-amplitude pulsating fontains, J. Fluid Mech. **366,** 333-350 (1998)

[47] D. de la Croix and O.Licandro, Life expectancy and endogenous growth, Econ. Lett. **65,** 255-263 (1999)

[48] M.C. Cross and P.C. Hohenberg, Pattern formation outside of equilibriuym, Rev. Mod. Phys. **65,** 851-1112 (1993)

[49] M.A. Dahlem, G. Hiller, A. Panchuk, and E. Schöll, Dynamics of delay-coupled excitable neural systems, Int. J. Bif. Chaos, in press (2008)

[50] J.-P. Dalmont, J. Gilbert, J. Kergomard, and S. Ollivier, An analytical prediction of the oscillation and extinction threshold of a clarinet, J. Acoust. Soc. Am. **118**, 3294-3305 (2005)

[51] O. Diekmann, S.A. van Gils, S.M. Verduyn Lunel, and H.-O. Walther, Delay Equations, Functional-, Complex-, and Nonlinear Analysis, Appl. Math. Sciences **110**, Springer, New York (1995)

[52] R.D. Driver, Ordinary and Delay Differential Equations, Applied Mathematical Sciences **20**, Springer-Verlag, New York (1977)

[53] J.C. Dunlap, Molecular bases for circadian clocks, Cell **96**, 271-290 (1999)

[54] I.R. Epstein and Y. Luo, Differential delay equations in chemical kinetics. Nonlinear models: The cross-shaped phase diagram of the Oregonator, J. Chem. Phys. **95**, 244-254 (1991)

[55] I.R. Epstein and J.A. Pojman, An Introduction to Nonlinear Chemical Dynamics, Oxford University Press, Oxford (1998)

[56] A. Erdélyi, A delay differential equation model of oscillations of exhange rates, Ph.D. Thesis Bratislava 2003

[57] T. Erneux, F. Rogister, A. Gavrielides, and V. Kovanis, Bifurcation to mixed external cavity mode solutions for semiconductor lasers subject to optical feedback, Opt. Comm. **183**, 467-477 (2000)

[58] T. Erneux, Asymptotic methods applied to semiconductor laser models, in Physics and Simulations of Optoelectronic Devices VIII, R. Binder, P. Blood, M. Osinski, Eds., Proc. SPIE **3944**, 588-601 (2000)

[59] T. Erneux, A. Gavrielides, and M. Sciamanna, Stable microwave oscillations due to external-cavity-mode beating in laser diodes subject to optical feedback, Phys. Rev. A **66**, 033809 (2002)

[60] T. Erneux and T. Kalmár-Nagy, Nonlinear stability of a delayed feedback controlled container crane, J. Vib. Cont. **13**, 606-616 (2007)

[61] T. Erneux, L. Larger, M.W. Lee, and J.P. Goedgebuer, Ikeda Hopf bifurcation revisited, Physica D **194**, 49-64 (2004)

[62] T. Erneux and H.-O. Walther, Bifurcation to large period oscillations in physical systems controlled by delay, Phys. Rev. E **72**, 066206 (2005)

[63] C.P. Fall, E.S. Marland, J.M. Wagner, and J.J. Tyson, Eds. Computational Cell Biology, Springer-Verlag, New York (2002)

[64] I. Fischer, R. Vicente, J.M. Buldú, M. Peil, C.R. Mirasso, M.C. Torrent, and J. García-Ojalvo, Zero-Lag Long-Range Synchronization via Dynamical Relaying, Phys. Rev. Lett. **97**, 123902 (2006)

[65] A.C. Fowler, Mathematical Models in the Applied Sciences, Cambridge Texts in Applied Mathematics, Cambridge University Press, Cambridge (1997)

[66] A.C. Fowler, An asymptotic analysis of the delayed logistic equation when the delay is large, IMA J. of Appl. Mathematics **28**, 41-49 (1982)

[67] A.C. Fowler, Asymptotic methods for delay equations, J. Eng. Math. **53**, 271-290 (2005)

[68] D.P. Francis, K. Wilson, L. C. Davies, A.J.S. Coates, and M. Piepoli, Quantitative general theory for periodic breathing in chronic heart failure and its clinical implications, Circulation **102**, 2214-2221 (2000)

[69] G.F. Gause, The Struggle for Existence, Williams and Wilkins, Baltimore (1934)

[70] A. Gavrielides and D.W. Sukow, Experimental observations, in Unlocking Dynamical Diversity: Optical Effects on Semiconductor Lasers, D.M. Kane and K.A. Shore Eds., J. Wiley & Sons Ltd, Chichester, UK (2005).

[71] N. Gavriely and O. Jensen, Theory and measurements of snores, J. Appl. Physiol. **74**, 2828-2837 (1993)

[72] G. Giacomelli, R. Meucci, A. Politi, and F.T. Arecchi, Defects and spacelike properties of delayed dynamical systems, Phys. Rev. Lett. **73**, 1099-1102 (1994)

[73] G. Giacomelli and A. Politi, Multiple scale analysis of delayed dynamical systems, Physica D **117**, 26-42 (1998)

[74] J.-P. Goedgebuer, M. Li, and H. Porte, Demonstration of bistability and multistability in wavelength with a hybrid acousto-optic device, IEEE J. Quant. Electron. **23**, 153-157 (1987)

[75] J.-P. Goedgebuer, L. Larger, H. Porte, and F. Delorme, Chaos in wavelength with feedback tunable laser diode, Phys. Rev. E **57**, 2795-2798 (1998)

[76] W. Goffmam and G. Harman, Mathematical approach to the prediction of scientific discovery, Nature **229 (5280)**, 103-104 (1971)

[77] A. Goldbeter, Computational approaches to cellular rhythms, Nature **420**, 238-245 (2002)

[78] M. Golubitsky and P.H. Rabinowitz. A sketch of the Hopf bifurcation theorem. In: Selected Works of Eberhard Hopf with Commentaries. C.S. Morawetz, J.B. Serrin and Y.G. Sinai, Eds., Amer. Math. Soc., Providence, 111-118 (2002)

[79] J. Grasman, Asymptotic Methods of Relaxation Oscillations and Applications, Applied Mathematical Sciences **63**, Springer, New York (1987)

[80] E.V. Grigorieva, H. Haken, and S.A. Kaschenko, Theory of quasiperiodicity in model of lasers with delayed optoelectronic feedback, Opt. Comm. **165**, 279-292 (1999)

[81] C. Grotta-Ragazzo, K. Pakdaman, and C.P. Malta, Metastability for delayed differential equations, Phys. Rev. E **60**, 6230-6233 (1999)

[82] W.S.C. Gurney, S.P. Blythe, and R.M. Nisbet, Nicholson's blowflies revisited, Nature **287**, 17-21 (1980)

[83] R. Haberman, Mathematical Models, Prentice-Hall, Englewood Cliffs, NJ (1977); reprinted by SIAM, Philadelphia (1998)

[84] J.K. Hale, Theory of Functional Differential Equations, Springer Verlag, New York (1977)

[85] J.K. Hale and S.M. Verduyn Lunel, Introduction to Functional Differential Equations, Appl. Math. Sciences **99**, Springer, New York (1993)

[86] T. Heil, I. Fischer, W. Elsässer, J. Mulet, and C.R. Mirasso, Chaos Synchronization and Spontaneous Symmetry-Breaking in Symmetrically Delay-Coupled Semiconductor Lasers, Phys. Rev. lett. **86**, 795-798 (2001)

[87] T. Heil, I. Fischer, W. Elsässer, and A. Gavrielides, Dynamics of semiconductor lasers subject to delayed optical feedback: The short cavity regime, Phys. Rev. Letters **87**, 243901 (2001)

[88] T. Heil, I. Fischer, and W. Elsässer, Stabilization of feedback-induced instabilities in semiconductor lasers, J. Opt. B: Quantum Semiclass. Opt. **2**, 413-420 (2000)

[89] V. Henri, Recherches sur la loi de l'action de la sucrase, C. R. Hebd. Acad. Sci. **133**, 891-894 (1901)

[90] R.J. Henry, Z.N.Masoud, A.H. Nayfeh, and D.T. Mook, Cargo pendulation reduction on ship-mounted cranes via boom-lu angle actuation, J. Vibration Control, **7**, 1253-1264 (2001)

[91] C. Hermite, Sur le nombre des racines d'une équation algébrique comprise entre des limites données, J. Reine Angew. Math. **52**, 39-51 (1856)

[92] R. Herrero, M. Figueras, J. Rius, F. Pi, and G. Orriols, Experimental observation of the amplitude death effect in two coupled nonlinear oscillators, Phys. Rev. Lett. **84**, 5312-5315 (2000)

[93] R. Herman, E.W. Montroll, R.B. Potts, and R.W. Rothery, Traffic dynamics: analysis of stability in car following, Oper. Res. E **17**, 86-106 (1958)

[94] R. Hinch and S. Schnell, Mechanism equivalence in enzyme-substrate reactions: Distributed differential delay in enzyme kinetics, J. Math. Chem. **35**, 253-264 (2004)

[95] H. Hirata, S. Yoshiura, T. Ohtsuka, Y. Bessho, T. Harada, K. Yoshikawa, R. Kageyama, Oscillatory expressio of the bHLH factor Hes1 regulated by a negative feedback loop, Science **298**, 840-842 (2002)

[96] A. Hjelmfelt and J. Ross, Experimental stabilization of unstable steady states in oscillatory and excitable reaction systems, J. Phys. Chem. **98**, 1176-1179 (1994)

[97] A. Hoffmann, A. Levchenko, M.L. Scott, and D. Baltimore, The IκB-NF-κB signaling module: Temporal control and selective gene activation, Science **298**, 1241-1245 (2002)

[98] A. Hohl and A. Gavrielides, Bifurcation cascade in a semiconductor laser subject to optical feedback, Phys. Rev. Lett. **82**, 1148-1151 (1999)

[99] Z. Hong, Z. Feizhou, Y. Jie, and W. Yinghai, Nonlinear differential delay equations using the Poincaré section technique, Phys. Rev. E **54**, 6925-6928 (1996)

[100] E. Hopf, Abzweigung einer periodischen Lösung von einer stationären Lösung eines Differentialsystems, Berichte Math.-Phys. Kl. Sächs. Akad. Wiss. Leipzig **94**, 3-32 (1942). A translation of this paper is in: J.E. Marsden and M. McCraken, The Hopf bifurcation and its applications, Appl. Math. Sci.**19**, Springer-Verlag, New York, 1976.

[101] J. Houlihan, D. Goulding, Th. Busch, C. Masoller, and G. Huyet, Experimental investigation of a bistable system in the presence of noise and delay, Phys. Rev. Lett. **92**, 050601 (2004)

[102] P. Hövel and E. Schöll, Control of unstable steady states by time-delayed feedback methods, Phys. Rev. E **72**, 046203 (2005)

[103] L.X. Huang and J.F.E. Williams, Neuromechanical interaction in human snoring and upper airway obstruction. J. Appl. Physiol. **86**, 1759-1763 (1999)

[104] Hurwitz, A., Uber die Bedingungen, unter welchen eine Gleichung nur Wurzeln mit negativen reellen Theilen besitz, Math. Ann. **46**, 273-284 (1895). English translation: On the conditions under which an equation has only roots with negative real part, in Selected papers on Mathematical Trends in Control Theory, ed. by R. Bellman and R. Kalaba, Dover, New York, 72-82 (1964)

[105] G.E. Hutchinson, Circular causal systems in ecology, Ann. New York Acad. Sci. **50**, 221-246 (1948)

[106] K. Ikeda, Multiple-valued stationary state and its instability of the transmitted light by a ring cavity system, Opt. Comm. **30**, 257-261 (1979); the derivation of Ikeda equation is simpler if we start from the Maxwell-Debye equations for highly dispersive media [107]; see [175], p 122; [180], p 39.

[107] K. Ikeda, H. Daido, and O. Akimoto, Optical turbulence: Chaotic behavior of transmitted light from a ring cavity, Phys. Rev. Lett. **45**, 709-712 (1980)

[108] K. Ikeda and O. Akimoto, Instability leading to periodic and chaotic self-pulsations in a bistable optical cavity, Phys. Rev. Lett. **48**, 617-620 (1982)

[109] L. Illing and D.J. Gauthier, Hopf bifurcations in time-delay systems with band-limited feedback, Physica D 210, 180 (2005)

[110] T. Insperger and G. Stépán, Stability improvements of robot control by periodic variations of the gain parameters, Proc. of the 11th World Congress in Mechanism and Machine science, Ed. Tian Huang, China Machinery Press, 1816-1820 (2004)

[111] G. Iooss and D.D. Joseph, Elementary stability and bifurcation theory, Springer, New York, 2nd Edition (1990)

[112] F. Ishida and Y.E. Sawada, Human hand moves proactively to the external stimulus: an evolutional strategy for minimizing transient error, Phys. Rev. Lett. **93**, 168105 (2004); Semianalytical transient

solution of a delayed differential equation and its application to the tracking motion in the sensory-motor system, Phys. Rev. E **75**, 012901 (2007)

[113] D. Israelsson and A. Johnsson, A theory for circumnutations in Helianthus annuus. Physiol. Plant. **20**, 957-976 (1967)

[114] M.H. Jensen, K. Sneppen, and G. Tiana, Sustained oscillations and time delays in gene expression of protein Hes1, FEBS Lett. **541**, 176-177 (2003)

[115] M.A. Johnson and F.C. Moon, Experimental characterisation of quasiperiodicity and chaos in a mechanical system with delay, Int. J. Bifurcation Chaos **9**, 49-65 (1999)

[116] A. Johnsson, Circumnutation: Results from recent experiments on earth and in space, Planta **203**, S147-S158 (1997)

[117] A. Johnsson, Geotropic responses in Helianthus and their dependence on the auxin ratio, Physiol. Plant **24**, 419-425 (1971)

[118] M. Kalecki, A macrodynamic theory of business cycle, Econometrica **3**, 327-344 (1935)

[119] V. Kolmanovskii and A. Myshkis, Introduction to the Theory and Applications of Functional Differential Equations, Mathematics and its Applications 463, Springer, New York (1999)

[120] T. Kalmár-Nagy, G. Stépán, and F.C. Moon, Subcritical Hopf bifurcation in the delay equation model for machine tool vibrations, Nonlinear Dynamics **26**, 121-142 (2001)

[121] T. Kalmár-Nagy, Delay-differential models of cutting tool dynamics with nonlinear and mode-coupling effects, Ph.D. Thesis Cornell University (2002)

[122] J. Keener and J. Sneyd, Mathematical Physiology, Springer-Verlag, New York (1998)

[123] B.E. Kendall, C.L. Briggs, W.W. Murdoch, P. Turchin, S.P. Ellners, E. McGauley, R.M. Nisbet, and S.N. Wood, Why do populations cycle? A synthesis of statistical and mechanistic modeling approaches, Ecology **80**, 1789-1805 (1999); available at: www2.bren.ucsb.edu/~kendall/pubs/1999Ecology.pdf.

[124] J. Kevorkian and J.D. Cole, Perturbation Methods in Applied Mathematics, Appl. Math. Sciences 34, Springer, New York (1981); Multiple Scale and Singular Perturbation Methods, Appl. Math. Sciences 114, Springer, New York (1996)

[125] S.E. Kingsland, Modeling Nature: Episodes in the History of Population Ecology, University of Chicago Press, Chicago (1985)

[126] I.Z. Kiss, Z. Kazsu, and V. Gáspár, Tracking unstable steady states and periodic orbits of oscillatory and chaotic electrochemical systems using delayed feedback control, Chaos **16**, 033109 (2006)

[127] L.E. Kollár and J. Turi, Numerical stability analysis in respiratory control systems models, Elect. J. Diff. Equations, Conference **12**, 65-78 (2005)

[128] Y.C. Kouomou, P. Colet, L. Larger, and N. Gastaud, Chaotic breathers in delayed electro-optical systems, Phys. Rev. Lett. **95**, 203903 (2005)

[129] Y.C. Kouomou, L. Larger, H. Tavernier, R. Bendoula, E. Rubiola, and P. Colet, Dynamic instabilities of microwaves generated with optoelectronic oscillators, Opt. Lett. **32**, 2571-2573 (2007)

[130] A. Krasovskii, Stability of Motion Translation, Stanford University Press, Stanford, CA (1963)

[131] N.M. Krylov and N.N. Bogolyubov, Introduction to Nonlinear Oscillations, translated from Russian by S. Lefschetz, Princeton University Press, Princeton, NJ (1943)

[132] Y. Kuang, Delay Differential Equations with Applications in Population Dynamics, Mathematics in Science and Engineering **191**, Academic Press, New York (1993)

[133] Y.A. Kuznetsov, Elements of Applied Bifurcation Theory, Appl. Mathematical Sciences **112**, Third Edition, Springer, New York(2004)

[134] E. Lacot, R. Day, and F. Stoeckel, Laser optical feedback tomography, Opt. Lett. **24**, 744-746 (1999)

[135] E. Lacot, R. Day, J. Pinel, and F. Stoeckel, Laser relaxation-oscillation frequency imaging, Opt. Lett. **26**, 1483-1485 (2001)

[136] R. Lang and K. Kobayashi, External optical feedback effects on semiconductor injection laser properties, IEEE J. Quant. Electr. **QE-16**, 347-355 (1980)

[137] J.P. Laplante, Stabilization of unstable states in hte bistable iodate-arsenous acid reaction in a continuous flow stirred tank reactor, J. Phys. Chem. **93**, 3882-3885 (1989)

[138] L. Larger, J.-P. Goedgebuer, and J.-M. Merolla, Chaotic oscillator in wavelength: A new setup for investigating differential difference equations describing nonlinear dynamics, IEEE J. Quant. Electron. **34**, 594-601 (1998)

[139] L. Larger, M.W. Lee, J.-P. Goedgebuer, W. Elflein, and T. Erneux, Chaos in coherence modulation: Bifurcations of an oscillator generating optical delay fluctuations, JOSA B **18**, 1063-1068 (2001)

[140] A. Lekebusch, A. Förster, and F.W. Scheinder, Chaos control in an enzymatic reaction, J. Phys. Chem. **99**, 681-686 (1995)

[141] J. Lewis, Autoinhibition with transcriptional delay: a simple mechanism for the zebrafish somitogenesis oscillator, Curr. Biol. **13**, 1398-1408 (2003)

[142] F. Lin and J.-M. Liu, Nonlinear dynamics of a semiconductor laser with delayed negative optoelectonic feedback, IEEE J. Quant. Electron. **39**, 562-568 (2003)

[143] A.J. Lotka, Elements of Physical Biology, Williams and Wilkins, Baltimore; reprinted as Elements of Mathematical Biology, Dover, New York (1956)

[144] A. Longtin and J.G. Milton, Modelling autonomous oscillations in the human pupil light reflex using non-linear delay-differential equations, Bull. Math. Biol. **51**, 605-624 (1989)

[145] A. Makroglou, J. Li, and Y. Kuang: Mathematical models and software tools for the glucose-insulin regulatory system and diabetes: An overview, Appl. Num. Math. **56**, 559-573 (2006)

[146] N. MacDonald, Biological Delay Systems: Linear Stability Theory, Cambridge University Press, New York, (1989)

[147] M.C. Mackey. Mathematical models of hematopoietic cell replication and control, pp. 149-178 in The Art of Mathematical Modelling: Case Studies in Ecology, Physiology and Biofluids (H.G. Othmer, F.R. Adler, M.A. Lewis, and J.C. Dallon eds.) Prentice-Hall, Englewood Cliffs, NJ (1997)

[148] M.C. Mackey and L. Glass, Oscillations and chaos in physiological control systems, Science **197**, 287-289 (1977)

[149] M.E. McIntyre, R.T. Schumacher, and J. Woodhouse, On the oscillations of musical instruments, J. Acoust. Soc. Am. **74**, 1325-1345 (1983)

[150] P. Mandel, Theoretical Problems in Cavity Nonlinear Optics, Cambridge Studies in Modern Optics, Camb. University Press, Cambridge (1997)

[151] P. Mandel and R. Kapral, Subharmonic and chaotic bifurcation structure in optical bistability, Optics Comm. **47**, 151-156 (1983)

[152] J. Mawhin, The legacy of Pierre-François Verhulst and Vito Volterra in population dynamics, To appear in The First 60 Years of Nonlinear Analysis of Jean Mawhin, Edited by M. Delgado, A. Suárez , J. López-Gómez, and R. Ortega, World Scientific Publishing, River Edge, NJ (2004); J. Mawhin, Les héritiers de Pierre-François Verhulst: une population dynamique, Bull. de la Classe des Sciences **7-12**, 349-378 (2002).

[153] R.M. May, Stability and Complexity in Model Ecosystems (2nd edition). Princeton University Press, Princeton (1975)

[154] R.M. May, Models for single populations pp 5–29, in R.M. May (ed.) Theoretical Ecology: Principles and Applications (2nd edition), Sinauer, Sunderland, MA (1981)

[155] C. Masoller, Distribution of residence times of time-delayed bistable systems driven by noise, Phys. Rev. Lett. **90**, 020601 (2003)

[156] Z.N. Masoud, A.H. Nayfeh, and A. Al-Mousa, Delayed position-feedback controller for the reduction of payload pendulations of rotary cranes, J. Vibration Control **9**, 257-277 (2003)

[157] Z.N. Masoud, A.H. Nayfeh, and D.T. Mook, Cargo pendulation reduction of ship-mounted cranes, Nonlinear Dynamics **35**, 299-311.(2004)

[158] Z.N. Masoud, A.H. Nayfeh, and N.A. Nafeh, Sway reduction on quayside container cranes using delayed feedback controller: Simulations and experiments, J. Vibration Control **11**, 1103-1122 (2005)

[159] J.C. Maxwell, On governors, Proc. Royal Soc. London **16**, 270-283 (1868)

[160] L. Michaelis and M.L. Menten, Die Kinetik der Invertinwerkung. Biochemische Zeitschrift **49**, 333 (1913) [as translated and excerpted in Mikulás Teich, A Documentary History of Biochemistry, 1770-1940, Fairleigh Dickinson University Press, Rutherford, NJ (1992)]

[161] W. Michiels and S.-I. Niculescu, Stability and Stabilization of Time-Delay Systems: an Eigenvalue-Based Approach, Adv. in Design and Control **12** SIAM, Philadelphia (2007)

[162] J.G. Milton, A. Longtin, A. Beuter, M.C. Mackey, and L. Glass, Complex dynamics and bifurcations in neurology, J. Theor. Biol. **138**, 129-147 (1989)

[163] J.G. Milton and A. Longtin, Evaluation of pupil constriction and dilation from cycling measurements, Vision Res. **30**, 515-525 (1990)

[164] N. Minorsky, Nonlinear Oscillations, D. Van Nostrand, Co., Inc., Princeton, NJ (1962), see Chapter 21

[165] N. Minorsky, Directional stability of automatic steered bodies, J. Amer. Soc. of Naval Engineers **34**, 280 (1922)

[166] N. Minorsky, Experiments with activated tanks, Trans. ASME 69, 735 (1947); Self-excited mechanical oscillations, J. Appl. Phys. **19**, 332 (1948)

[167] A.D. Mishkis, Linear Differential Equations with Retarded Arguments, Nauka, Moscow (1972)

[168] D.G. Mitchell and D.H. Klyde, Identifying a PIO signature - New techniques applied to an old problem, AIAA Atm. Flight Mech. Conf., Keystone, Colorado (2006)

[169] N.A.M. Monk, Oscillatory expression of Hes1, p53, and NF-κB driven by transcriptional time delays, Curr. Biol. **13**, 1409-1413 (2003)

[170] J. Mork, B. Tromborg, and P.L. Christiansen, Bistability and low frequency fluctuations in semiconductor lasers with optical feedback: A theoretical analysis, IEEE J. Quantum Electron. **24**, 123-133 (1988)

[171] J. Mork and B. Tromborg, The mechanism of mode selections for an external cavity laser, IEEE Photon. Technol. Lett. **2**, 21-23 (1990)

[172] J.D. Murray, Mathematical Biology I: An Introduction, Inter. Appl. Mathematics **17**, Springer, Berlin, Third Edition (2002)

[173] P. Nardone, P. Mandel, and R. Kapral, Analysis of a delay-differential equation in optical bistability, Phys. Rev. A **33**, 2465-2471 (1986)

[174] N.A. Nayfeh, Adaptation of delayed position feedback on the reduction of sway of container cranes, Master's thesis, Virginia Polytechnic Institute and State University (2002)

[175] A.C Newell and J.V. Moloney, Nonlinear Optics, Addison-Wesley, New York (1992)

[176] A.J. Nicholson, The self-adjustment of populations to change, Cold Spring Harbor Symposium on Quantitative Biology **22**, 153–173 (1957)

[177] E. Niebur, H. G. Schuster, and D. Kammen, Collective frequencies and metastability in networks of limit-cycle oscillators with time delay, Phys. Rev. Lett. **67**, 2753-2756 (1991)

[178] M. Nizette, Front dynamics in a delayed-feedback system with external forcing, Physica D **183**, 220-244 (2003)

[179] H. Olesen, J.H. Osmundsen, and B. Tromborg, Nonlinear dynamics and spectral behavior for an external cavity laser, IEEE J. Quantum Electron. 22, 762-773 (1986)

[180] K. Otsuka, Nonlinear Dynamics in Optical Complex Systems, KTK Scientific/Tokyo, Kluwer Academic, Boston (1999)

[181] H. Park, K.-S. Hong, Boundary control of container cranes, Proc. SPIE, Vol. 6042, 604210 (2005)

[182] P. Parmananda, R. Madrigal, M. Rivera, L. Nyikos, I.Z. Kiss, and V. Gáspár, Stabilization of unstable steady states and periodic orbits in an electrochemical system using delayed-feedback control, Phys. Rev. E **59**, 5266-5271 (1999)

[183] P. Parmananda, Tracking fixed-point dynamics in an electrochemical system using delayed-feedback control, Phys. Rev. E **67**, 045202 (2003)

[184] R. Pearl and L.J. Reed, On the rate of growth of the population of the United States since 1790 and its mathematical representation, Proc. Nat. Acad. Sci. USA **6**, 275-288 (1920)

[185] D. Perlmutter, Stability of Chemical Reactors, Prentice-Hall, Englewood Cliffs, NJ (1972)

[186] R.J. Peterka, Sensorimotor integration in human postural control, J. Neurophysiol. **88**, 1097-1118 (2002)

[187] R.J. Peterka and P.J. Loughlin, Dynamic regulation of sensorimotor integration in human postural control, J. Neurophysiol **91**: 410-423 (2004)

[188] D. Pieroux, T. Erneux, and K. Otsuka, minimal model of a class-B laser with delayed feedback: Cascading branching of periodic solutions and period-doubling bifurcation, Phys. Rev. A **50**, 1822-1829 (1994)

[189] D. Pieroux, T. Erneux, A. Gavrielides, and V. Kovanis, Hopf bifurcation subject to a large delay in a laser system, SIAM J. Appl. Math. **61**, 966–982 (2000)

[190] D. Pieroux, T. Erneux, B. Haegeman, K. Engelborghs, and D. Roose, Bridges of periodic solutions and tori in semiconductor lasers subject to delay, Phys. Rev. Lett. **87**, 193901 (2001)

[191] G.D. Pinna, R. Maestri, A. Mortara, M.T. La Rovere, F. Fanfulla, and P. Sleight, Periodic breathing in heart failure patients: testing the hypothesis of instability of the chemoreflex loop, J. Appl. Physiology **89**, 2147-2157 (2000)

[192] E. Pinney, Ordinary Difference-Differential Equations, University of California Press, Berkeley (1958)

[193] O. Pourquie, The segmentation clock: Converting embryonic time into spatial pattern, Science **301**, 328-330 (2003)

[194] K. Pyragas, Control of chaos via extended delay feedback, Phys. Lett. A **206**, 323-330 (1995)

[195] K. Pyragas, V. Pyragas, I.Z. Kiss, and J.L. Hudson, Stabilizing and tracking unknown steady states of dynamical systems, Phys. Rev. Lett. **89**, 244103 (2002)

[196] D.V. Ramana Reddy, A. Sen, and G.L. Johnston, Time delay effects on coupled limit-cycle oscillators at Hopf bifurcation, Physica D **129**, 15-34 (1999)

[197] D.V. Ramana Reddy, A. Sen, and G.L. Johnston, Time delay induced death in coupled limit cycles oscillators, Phys. Rev. Lett. **80**, 5109-5112 (1998)

[198] D.V. Ramana Reddy, A. Sen, and G.L. Johnston, Experimental evidence of time-delay-induced death in coupled limit-cycle oscillators, Phys. Rev. Lett. **85**, 3381-3384 (2000)

[199] B.F. Redmond, V. G. LeBlanc, and A. Longtin, Bifurcation analysis of a class of first-order nonlinear delay-differential equations with reflectional symmetry, Physica D **166**, 131-146 (2002)

[200] J.V. Ringwood and S. V. Malpas, Slow oscillations in blood pressure via a nonlinear feedback model, Am. J. Regulatory Integrative Comp Physiol **280**: R1105-R1115 (2001)

[201] E.J. Routh, A Treatise on the Stability of a Given State of Motion - Adams Prize Essay, Macmillan, New York (1877)

[202] M.R. Roussel, The use of delay differential equations in chemical kinetics, J. Phys. Chem. **100**, 8323-8330 (1996)

[203] P. Saboureau, J.P. Foing, and P. Schanne, Injection-locked semiconductor lasers with delayed optoelectronic feedback, IEEE J. Quantum Electron. **33**, 1582-1591 (1997)

[204] T. Sano, Antimode dynamics and chaotic itinerancy in the coherence collapse of semiconductor lasers with optical feedback, Phys. Rev. A **50**, 2719-2726 (1994)

[205] J.M. Saucedo Solorio, D.W. Sukow, D.R. Hiks, and A. Gavrielides, Bifurcations in a semiconductor laser subject to delayed incoherent feedback, Opt. Comm. **214**, 327-334 (2002)

[206] M. Sciamanna, K. Panajotov, H. Thienpont, I. Veretennicoff, P. Mégret, and M. Blondel, Optical feedback induces polarization mode hopping in vertical-cavity surface-emitting lasers, Opt. Lett. **28**, 1543-1545 (2003)

[207] M. Schanz and A. Pelster, Analytical and numerical investigations of the phase-locked loop with time delay, Phys. Rev. E **67**, 056205 (2003)

[208] V.B. Scheffer, The rise and fall of a reindeer herd, Scientific Monthly, **73**, 356–362 (1951)

[209] F.W. Schneider, R. Blittersdorf, A. Förster, T. Hauck, D. Lebender, and J. Müller, Continuous control of chemical chaos by time delayed feedback, J. Phys. Chem. **97**, 12244-12248 (1993)

[210] Handbook of Chaos Control, E. Schöll and H.G. Schuster (Eds.), Wiley-VCH, 2nd Rev. Enl. Edition (2007)

[211] N. Schunk and K. Petermann, Numerical analysis of the feedback regimes for a single-mode semiconductor laser with external feedback, IEEE J. Quantum Electron. **24**, 1242-1247 (1988)

[212] H.G. Schuster and P. Wagner, Mutual entrainment of two limit cycle oscillators with time delayed coupling, Prog. Theor. Physics **81**, 939-945 (1989)

[213] J. Sieber, Numerical bifurcation analysis for multi-section semiconductor lasers. SIAM J. Appl. Dyn. Sys. **1**, 248-270 (2002)

[214] H.M. Shi and S.A. Tobias, Theory of finite amplitude machine tool instability, Int. J. Mach. Tool Des. Res. **24**, 45-49 (1984)

[215] J.E. Socolar, D.W. Sukow, and D. J. Gauthier, Stabilizing unstable periodic orbits in fast dynamical systems, Phys. Rev. E **50**, 3245-3248 (1994)

[216] A.S. Somolinos, Periodic solutions of the sunflower equation: $x'' + (a/r)x' + (b/r)\sin x(t-r) = 0$, Quart. Appl. Math. **35**, 465-478 (1978)

[217] G. Stépán, Retarded Dynamical Systems, Longman, London (1989)

[218] G. Stépán and L. Kollár, Balancing with reflex delay, Math.Comput. Model. **31**, 199-205 (2000)

[219] S.H. Strogatz, Nonlinear Dynamics and Chaos: With Applications in Physics, Biology, Chemistry, and Engineering, Addison-Wesley, Reading, MA (1994)

[220] D.W. Sukow, M.C. Hegg, J.L. Wright, and A. Gavrielides, Mixed external cavity mode dynamics in a semiconductor laser, Opt. Lett. **27**, 827-829 (2002)

[221] M. Szydlowski and A. Krawiec, Scientific cycle model with delay, Scientometrics **52**, 83-95 (2001)

[222] A.A. Tager and B.B. Elenkrig, Stability regimes and high-frequency modulation of laser diodes with short external cavity, IEEE J. Quant. Electron. **29**, 2886-2890 (1993)

[223] A.A. Tager and K. Petermann, High-frequency oscillations and self-mode locking in short external cavity laser diodes, IEEE J. Quant. Electron. **30**, 1553-1561 (1994)

[224] A. Takamatsu, T. Fujii, and I. Endo, Time delay effect in a living coupled oscillator system with the plasmodium of *physarum polycephalum,* Phys. Rev. Lett. **85**, 2026-2029 (2000)

[225] S. Tang and J.M. Liu, Chaotic pulsing and quasi-periodic route to chaos in a semiconductor laser with delayed opto-electronic feedback, IEEE J. Quant. Electron. **37**, 329-336 (2001)

[226] S. Tang, R. Vicente, M.C. Chiang, C.R. Mirasso, and J.-M. Liu, Non-linear dynamics of semiconductor lasers with mutual optoelectronic coupling, J. Select. Topics Quantum Electron. **10**, 936-943 (2004)

[227] S. Thompson Delay-differential equations. Scholarpedia, **2**(3):2367 (2007)

[228] L.S. Tsimring and A. Pikovsky, Noice-induced dynamics in bistable systems with delay, Phys. Rev. Lett. **87**, 250602 (2001)

[229] L.S. Tuckerman and D. Barkley, Bifurcation analysis of the Eckhaus instability, Phys. D **46**, 57-86 (1990)

[230] J.J. Tyson, K. Chen, and B. Novak, Network dynamics and cell physiology, Nat. Rev. Mol. Cell Biol. **2**, 908-916 (2001)

[231] J.J. Tyson, Oscillations, bistability and echo waves in models of the Belousov-Zhabotinskii reaction, Ann. N.Y. Acad. Sci. **316,** 279-295 (1979)

[232] G.H.M. van Tartwijk, A.M. Levine, and D. Lenstra, Sisyphus effect in semiconductor lasers with optical feedback, IEEE J. Select. Top. Quant. Electron. **1,** 466-472 (1995)

[233] V. Torre, Existence of limit cycles and control in complete Keynesian systems by theory of bifurcations, Econometrica **45,** 1457-1466 (1977)

[234] A. Uppal, W.H. Ray, and A.B. Poore, On the dynamic behavior of continuous stirred tank reactors, Chem. Eng. Sci. **29,** 967-985 (1974)

[235] O. Ushakov, S. Bauer, O. Brox, H.-J. Wünsche, and F. Henneberger, Self-organization in semiconductor lasers with ultrashort optical feedback, Phys. Rev. Lett. **92,** 043902 (2004)

[236] J.V. Uspenky, Theory of Equations, McGraw-Hill, New York (1948)

[237] P.F. Verhulst, Notice sur la loi que la population suit dans son accroissement, Correspondance mathématique et physique **10,** 113-125 (1838)

[238] E. Villermaux, Pulsed dynamics of fountains, Nature **371,** 24-25 (1994)

[239] E. Villermaux and E.J. Hopfinger, Self-sustained oscillations of a confined jet: a case study for the non-linear delayed saturation model, Physica D 72, 230-243 (1994)

[240] E. Villermaux, Memory-induced low frequency oscillations in closed convection boxes, Phys. Rev. Lett. **75,** 4618-4621 (1995)

[241] J. Weiner, F.W. Schneider, and K. Bar-Eli, Delayed-Output-Controlled Chemical Ocillations, J. Phys. Chem. 93, 2704-2711 (1989)

[242] G.B. Whitham, Linear and Nonlinear Waves, Wiley-Interscience, New York (1974)

[243] M. Wolfrum and D. Turaev, Instabilities of Lasers with moderately delayed optical feedback, Opt. Comm. **212,** 127-138 (2002)

[244] M. Wolfrum and S. Yanchuk, Eckhaus instability in systems with large delay, Phys. Rev. Lett. **96,** 220201 (2006)

[245] H.-J. Wünsche, S. Bauer, J. Kreissl, O. Ushakov, N. Korneyev, F. Henneberger, E. Wille, H. Erzgräber, M. Peil, W. Elsässer, and I. Fischer, Synchronization of delay-coupled oscillators: A study of semiconductor lasers, Phys. Rev. Lett. **94,** 163901 (2005)

[246] E.C. Zimmermann and J. Ross, Light induced bistability in $S_2O_6F_2 \rightleftharpoons 2\ SO_3F$: Theory and experiment, J. Chem. Phys. **80**, 720-729 (1984)

[247] E.C. Zimmermann, M. Schell, and J. Ross, Stabilization of unstable states and oscillatory phenomena in an illuminated thermochemical system: Theory and experiment, J. Chem. Phys. **81**, 1327-1336 (1984)

[248] D. Zwillinger, Handbook of Differential Equations, Academic Press, Boston (1989), p. 172

[249] http://www.theorem.net/theorem/lewis1.html, A brief history of feedback control, reprinted by permission from Chapter 1: Introduction to Modern Control Theory, in F.L. Lewis, Applied Optimal Control and Estimation, Prentice-Hall, Englewood Cliffs, NJ (1992)

[250] http://-groups.dcs.st-andrews.ac.uk/~history/Mathematicians

Index